Commonsense Statist
and Others

STUDENTS LIBRARY OF ECONOMICS

Commonsense Statistics for Economists and Others

F. R. Jolliffe
Brunel University

LONDON AND BOSTON
ROUTLEDGE & KEGAN PAUL

First published in 1974
by Routledge & Kegan Paul Ltd
Broadway House, 68–74 Carter Lane,
London EC4V 5EL and
9 Park Street,
Boston, Mass. 02108, USA
Set in Times Roman
and printed in Great Britain by
Willmer Brothers Limited, Birkenhead

ISBN 0 7100 7952 4 (c)
ISBN 0 7100 7953 2 (p)
Library of Congress Catalog Card No. 74–81313

General editor's introduction

It is apparent that, even at the most elementary level, it is necessary nowadays for students of the social sciences and, in particular, for economists, to have some serious acquaintance with quantitative methods. The problem has been to find for some of them a textbook which is sufficiently elementary but is also reasonably interesting. It seems to me that the present volume by Mrs Jolliffe fulfils these ends admirably. She is at great pains to explain all the relevant material with great care and quotes examples which will enable the student to master the relevant material. My own view is that the book will also be of interest to a great number of other students who wish to pursue a short course in statistics.

M.H.P.

To my son Godric who was born
half-way through

Contents

Preface

This book aims to teach the fundamental ideas of statistics in a manner that should not be scorned by the numerate and will be understood by the non-numerate. Algebraic manipulations and arithmetic are kept to a minimum, but reasons and proofs are indicated wherever possible. Little more than an ability to count and some commonsense are needed in order to understand the book, but familiarity with algebraic notation, substitution in formulae, the summation sign, and permutations and combinations, is a useful prerequisite. The more numerate reader should be able to fill in the gaps in the algebra for himself, or progress easily from this to more advanced texts.

For a knowledge of statistics to be at all useful it must include something on both the theoretical and applied sides. Thus, this book gives many examples using either real data where indicated or realistic imaginary data as illustrations of the methods. The book is designed to be read in the order written, but more mathematical readers might prefer to read chapters 2 and 3 after the other chapters (reference to chapters 2 and 3 is given in later chapters where needed). Throughout, the reader should try to separate definitions and methods of calculation in his mind. Although the book is aimed primarily at the economist this should not exclude other classes of reader.

I should like to thank the following for their help in reading and criticizing the first draft of the book – Professor M. H. Peston, Mrs H. J. Lamaison, my father Mr S. D. H. Savigear, and my husband Clive. Thanks are also due to Mr R. J. Allard and Mr S. Pollock for helpful suggestions, to my students at Southampton University and Queen Mary College who have contributed more than they realize, and to the secretaries at Queen Mary College for their willing typing. I am grateful to the Senate of the University of London for permission to use parts of examination questions in chapter 2 question 6, chapter 3 questions 1, 2, 6, 8, chapter 4 question 10, chapter 5 questions 3, 9, chapter 6 questions 7, 8, 10, and chapter 7 questions 1, 3, and 8.

I
About statistics

The beginning of statistical study occurs with the emergence of probability theory, which we can describe loosely as being theory relating to the chance or likelihood of events. In classical Greece and Rome elaborate games of dice were played. The best throw was called the 'venus'. The Roman emperors were keen players, and the Emperor Claudius wrote a book, unfortunately lost, on how to win at dice. In the temples, oracles gave divine interpretations to different combinations of throws of dice, each throw being given the name of a god. Thus it seems that by this time the relative frequencies with which different throws occurred had at least been observed, so that in dice games the biggest prizes could be allotted to the throws occurring least often, and in the temples oracles could adjust omens to fit in with the frequencies of occurrence they thought desirable.

In the mid-sixteenth century the Italian Girolamo Cardano wrote a book on games of chance, and this is the first known work calculating theoretical probabilities correctly. Another important development in probability theory was the famous correspondence, beginning in 1654, between the French mathematicians Fermat and Pascal, on a problem concerning dice posed by the gambler Chevalier de Méré. This relates to the fact that in playing with ordinary six-sided dice there is slightly more than half a chance of throwing one six with one die in four throws, and slightly less than half a chance of throwing two sixes with two dice in twenty four throws, as had been observed by de Méré, and as can be shown by elementary probability theory. The point that troubled de Méré was as follows. In the case of one die there are six possible results on a single throw (1, 2, 3, 4, 5, 6) and four throws are being discussed, suggesting there are four chances in six of getting one six. In the case of two dice thrown together there are thirty-six possible results on a single throw (any one of the six numbers on one die can appear with any of the six numbers on the other) and twenty-four throws are being discussed, suggesting there are twenty-four chances in thirty-six of getting two sixes. Since 4:6 = 24:36 de Méré argued that there is a fallacy in the theory of numbers. His calculations are in fact in error as the situation is less simple than he imagined.

Demography can be considered to be a branch of statistics and its beginnings occurred with the work of John Graunt in England in the seventeenth century. Parish registers contained details of numbers of christenings, weddings, and burials and Bills of Mortality contained details of numbers of deaths from different causes. Graunt used this and other information and one interesting calculation he did was to find the worst year of the Great Plague. For a full historical account of statistics from the earliest times to the Newtonian era see David (1962).

The word 'statistics' dates from the mid-eighteenth century when it was used to describe the study of the 'political arrangements of the modern states of the known world'. At first these studies were descriptive, but later took account of numerical data, and as methods developed the subject of statistics was extended into other fields. One of the first applications was by the Belgian Quetelet (1835–70) who applied probability theory to anthropological measurements.

Sir Francis Galton (1822–1911) knew of Quetelet's studies and was interested in heredity and classifying men by physical and other characteristics. His study on the heights of fathers and sons (do tall fathers have tall sons?) may be said to be the start of regression theory. Contemporary with Galton was Gregor Mendel (1822–84) who discovered laws of heredity by means of experiments on peas. Mendel's laws became the foundation for the theory of genetics which depends a great deal on probability theory.

In more recent times Sir R. A. Fisher (1890–1962) is accepted as one of the chief founders of modern statistics. Much of his work was done at the agricultural experimental station at Rothampstead and many of his examples relate to agriculture. He was concerned, amongst other things, with the design of experiments, that is, with designing efficient experiments which could be analysed statistically.

For examples of other applications of statistics developed largely in this century consider medicine, insurance, economics, and social science in general. In medicine for instance there are studies of the spread of diseases, and more recently studies of the possible connection between smoking and cancer. In insurance, which owes its origins to the seventeenth-century astronomer and mathematician Edmund Halley, probability theory is at the basis of the calculation of premiums, but some of the problems where statistics might help, for instance in motor insurance, are very complex and are as yet only partially solved. In economics attempts are being made to build models of the economy which aid understanding of the current economic situation and which will be of use in forecasting.

As regards applications of statistics in the social sciences in general, note that in this age of computers, facts are frequently presented in a statistical manner on television, the radio, and in newspapers, and many of these are of concern to social scientists. One has in mind such things as balance of payments figures, the Index of Retail Prices (commonly called the Cost of Living Index),[1] figures on voting, and figures concerning the effectiveness of new drugs. Parallel with this increase in presentation of facts is the growth of social science as an academic discipline, and the development of statistical methods of particular use in social science. It is almost true to say that no one can avoid quantitative matters of a statistical kind, and in a sense we must all become statisticians.

The word 'statistics' may mean a number of things. We must, in particular, distinguish between statistics in the plural and statistics in the singular. In the plural, statistics, especially in common usage, means a collection of figures, data, information. An example is vital statistics, whether in its meaning of a certain set of measurements of a woman's body, or its more serious meaning of figures relating to births, deaths, marriages, and other demographic data concerning areas such as countries. Other examples are statistics of prices, incomes, and productivity, education statistics, and crime statistics. All libraries of any size have collections of published statistics on these and many other topics of interest to general members of the community. (See Appendix 1 and Central Statistical Office, 1972.)

Statistics in the singular refers to the subject *statistics*. It is a set of methods used for making decisions where there is uncertainty arising from the incompleteness or instability of the information available. Both *statistical theory* and what may be called *applied statistics* can be seen as in some way falling under this heading. Thus, included here are methods of summarizing data, methods of estimation, and methods of sampling. But all these depend on such topics

[1] The Cost of Living Index was started in 1914 and was used until January 1956 when it was replaced by the Index of Retail Prices. See Department of Employment, *Family Expenditure Survey Reports*.

as probability theory and the theory of distributions. It is worth mentioning here that statistical theory and methods have developed to a large extent as a branch of mathematics, but also sometimes as a result of trial and error and the need for a quick answer in a laboratory or on a factory floor. There is a two way process sometimes starting from an extremely abstract theorem and working to a practical application, and sometimes starting from a real problem and working back to a theoretical generalization.

Theoretical statistics is almost a branch of pure mathematics, with all the rigorous proofs, symbolism, assumptions, and the niceties of algebra that that involves. *Applied statistics*, although it still has its technical difficulties, is less abstract. It is concerned with real problems and applies the theory to statistics (meaning data). In a sense, to the pure theorist the data as such, whether they come from the social sciences, the pure sciences, medicine, engineering or business, do not matter. In practice, of course, they do. In particular, applied methods are needed to provide answers fairly quickly and cheaply, both before and after collection of the data. Without applications there would be little point in the theory. Without theory there is no guide as to how to deal with the data, and no justification of any treatment of them.

Thus, theoretical and applied statistics complement one another. In this connection there is a word of caution to be uttered although it will not really affect the student at the start of his statistical studies. The assumptions so necessary to the theoretician rarely hold exactly in practice. The problem for the applied man, therefore, is to see that they hold well enough. He must not use rules crudely and automatically and he needs some understanding of the theories behind the methods in order not to misuse them. At the same time it is important not to fall into the erroneous view that the conditions never hold for the application of any methods. Of course, the theoretician who has experience of data and who knows the needs of those working with data will seek to develop methods where the assumptions are not too limiting.

Before going on to the technical parts of the book we describe some problems of a statistical nature which a fictional manufacturer meets. These serve as an introduction to what statistics is about. In addition books which give a good idea of what statistics is about without becoming too technical are Huff (1954); Kalton (1966); and Zeisel (1958).

Suppose a manufacturer wants to know the potential market for an improved model of a consumer durable and he decides to collect information on this by means of a sample survey. The product may be a washing machine, the potential market for which will be all the households in the country. It would, however, be prohibitively expensive to ask each household whether it would buy a particular machine and at what price, so the manufacturer might decide to ask questions of a selection of households: that is, take a sample.

In choosing this sample of households for his survey a number of questions arise. For example, how can a sample be chosen which has the characteristic that basically what is true of the sample is true of the population, or nearly true within some degree of error? Now, it is likely that views in the north of the country are different from those in the south. Can the manufacturer's sample take account of these regional differences and other differences between households, such as differences in income, and differences in the number of children contained in households, for instance? If he chooses a sample of persons by standing on a street corner will it include too many housewives and not enough households with working wives? Would he be better to choose names from a list such as the electoral register? How large a sample should he take?

In addition to deciding on the sampling method, the

manufacturer has to consider what questions he is going to ask, and how. This is a large topic, and not one which we can cover to any great extent in this book. Survey methodology is, however, an interesting and important subject. (Interested readers are referred to Moser and Kalton, 1971.) Let us suppose then that the survey is done and a wealth of data of various kinds has been collected. This is sometimes referred to as raw data, and by itself has little or no meaning or value. Moreover, it is not very illuminating reading through answers to questions, respondent by respondent. We are interested in any patterns there are to be found and possibly these will emerge if the answers can be grouped.

As a start, straight counts could be made, so that for instance, the age composition of the sample becomes apparent. Are older persons at all interested in possessing modern equipment? How does their interest compare with younger people? More generally a cross-tabulation is needed showing for example how many in each age group are interested. We can then similarly determine proportions such as the proportion of persons in the north of the country who already have the current version of the consumer durable, say.

The next step might be to look at age and region together. We might then need to know if the persons in the north of the country are on the whole older than those in the south. Is there a summary measure of age in each area which will show this and possibly remove the need for looking at the whole frequency distribution of ages? Can we also summarize the variability in the data and so show for instance that in one area persons all earn about the same amount, whereas in another there are several persons in all income groups from very high to very low?

The manufacturer feels that he will better be able to believe the survey results if he can check on his sampling procedure by seeing if his sample is a population in miniature.

Data on the population are published in the *Census of Population* (Office of Population Censuses and Surveys, 1961, 1966, 1971) which gives figures on many demographic, housing and employment variables. It is possible to test the extent to which differences between the distributions of these variables in the population and the sample matter by comparing the whole distributions, or summary measures, or proportions.

Similarly there may be differences between the north and the south, or between other sub-samples, or possibly among a set of sub-samples. Can the manufacturer conclude that there really is or is not a difference? How certain is he that his conclusion is right? Is there any possibility that he is wrong, and does he know the chance that he is wrong?

The original object of the survey was to assess the market for a particular good. What we have are figures for a sample. How can the demand in the sample be multiplied up to give the demand in the population and will this estimate be satisfactory? Perhaps the estimate would seem more reasonable if lower and upper limits could be given to it. What confidence do we have that such an interval estimate is correct? It is easy to see that from comparatively simple beginnings we are already involved with problems of some depth and complexity. But there is more!

The manufacturer thinks that the exercise would be more useful in the long run if he could discover what it is that determines spending power, because it is this which indirectly decides whether or not a person buys his product. He might start by comparing relationships between pairs of variables, such as age and spending power (to which we can suppose he is able to attach a numerical value), to see whether there is any connection or correlation between them. If an equation can be fitted to describe the relation, this might be useful for predictive purposes. At each stage knowing the confidence that could be placed in the result would be useful. As the study progresses relationships between whole sets of variables may be of interest, and possibly a model of, say, consumer spending will emerge.

While the survey is in progress other studies and developments take place. Sales data are produced in various forms. Monthly figures of sales seem particularly interesting, but the long term upward trend in sales is confused with seasonal effects and effects on sales caused by changes in taxation of various kinds. These different components need to be sorted out. In improving the product various processes are tried on small numbers of items to see if it is worth while using them in full-scale production. Reaching conclusions from such small numbers is tricky.

When all these different studies have been completed the manufacturer wants to present them in a printed report to the board of directors of the firm. He thinks the results will be easiest to understand if they are in diagram form. However, he suspects he needs also to give details of how the results were obtained and to present them in such a way that there is no doubt as to their validity.

This is just one example. The reader will be able to think of many more derived from his own experience. They ought to convince him of the need for theory, but, presumably, theory related to the problem in hand.

In the remainder of this book ways of dealing with most of the statistical questions presented in discussing our washing machine manufacturer will be given (although specific reference to them and him is not always made). Of course, all of this is relevant in a much wider context. Other problems not described as yet are also covered. More to the point, the reader is left to become a statistical consultant and fit problems that interest him into a statistical framework, the solution to which will be assisted by what he reads and learns here.

2

Descriptive statistics
Part I — Presentation of data

Data and measurement

The previous chapter should have made clear that in applied statistics we deal with information or data. As social scientists we might be interested in the types of housing different families occupy, incomes of heads of households, or trade figures of different countries. Possibly we have several pieces of information relating to the same person such as his income, age and education; or to a firm as product manufactured, the number of employees, the area of occupied floor space and the gross annual turnover. Thus as soon as we start thinking about data with a view to understanding what they mean we realize that there are different types of data.

At the lowest level of measurement items are identified just by names (which might be symbols or numbers). As examples take the sex of a person, or his country of birth, or the raw materials purchased by a firm. The measurement is said to be on a *nominal* scale. On this scale the choice of names is limitless, but the number of operations which can be performed on the data is small. We cannot do much more than count how many items there are in a group, and combine the groups.

When the groups, identified by names, have an ordering, as for example social class groups, or degree of severity of burns, or level of difficulty of a course of study, the measurement is said to be on an *ordinal* scale. In this case operations on the data have the limitation that they must preserve the ordering. (An ordering means that we can compare things according to the criteria more than, or less than, or equal to.)

When the distances between ordered groups are known the groups can be identified by meaningful numbers rather than by names. Any set of numbers such that the ordering and the relative distances between the groups are kept can be used. Measurement is said to be on an *interval* scale. A well-known example of a quantity measured on this scale is temperature, where the scales of Fahrenheit and of Centigrade are commonly used. In interval scales the unit of measurement and the position of the zero are arbitrary.

The highest level of measurement is on the *ratio* scale. This is like an interval scale with a true zero. Most of the things we measure by numbers are on this scale, for example, measurement of lengths, incomes, or the number of years lived at one's current permanent address. Notice that even when there is a choice of scales, as, say, measuring lengths in inches or in centimetres, the position of the zero is fixed. Clearly, far more can be done with measurements on this scale than with those on the nominal scale. Observe that we can change in the direction, ratio scale to interval scale to ordinal scale to nominal scale, but not the other way round.

It is sometimes useful to refer to the nominal and ordinal scales as 'qualitative', and to the interval and ratio scales as 'quantitative'.

Notice also that variables can be either *discrete* or *continuous*. Discrete means distinct, and discrete variables can take only a finite number of values. Examples of discrete variables are the number of children in a family, the number of absentees from work each week, and shoe sizes. In contrast, continuous variables can take any (real) value within a range of values, for example, age, yield of crops, and speed of cars on a motorway. Notice, however, that because of limitations in our measuring apparatus, readings of values of continuous variables are in a discrete form. In addition continuous variables appear to be discrete when values are rounded as, for example, with age given to the nearest year, when only whole numbers occur.

Frequency distributions (ungrouped data)

If we have qualitative data, a natural first step to take in summarizing the information contained in the data is to count up how many times each identifier appears. Similarly if we have discrete quantitative data and it is practical to do so, the corresponding operation is to count up how many times each value of the variable appears. In both cases we say in statistical language that we are forming a *frequency distribution*, or table. The number of times an identifier or value occurs is called its *frequency*.

There are a few points to notice about frequency distributions of discrete variables. We shall denote the variable by X, and the frequency of a particular X by $f(X)$. The total number of Xs (which is not in general the same as the number of different values taken by X) is called the total frequency and it is customary to denote this by n where $n = \sum f(X)$ (see Appendix 2). The ratio $f(X)/n$ is known as the *relative frequency* of X, and this quantity is important in probability theory. If we count up how many observations are less than or more than a particular value, the numbers obtained are called *cumulative frequencies*. They can be found by adding together appropriate frequencies $f(X)$ as shown in the example below.

As an example consider the discrete frequency distribution

in Table 2.1 which relates to new houses constructed in 1970 for local authorities and new towns in England and Wales (derived from *Social Trends No. 2* – see Government Statistical Service, 1971). In this case X is the number of bedrooms and takes values 0, 1, 2, 3, and so on.

Table 2.1

No. of bedrooms [X]	*Frequency* [$f(X)$]
1	5 670
2	15 120
3	38 430
4	2 520
5	1 260
	63 000

Notice that values of the variable are listed vertically in numerical order (here from lowest to highest, but sometimes highest to lowest is preferable), columns are labelled, and the total frequency of 63 000 is written in. If, instead of 1 260 houses with 5 bedrooms we had 1 200 with 5 and 60 with 6 bedrooms we might have a label under X of '5 or more' with the frequency of 1 260 rather than two separate labels '5' and '6'. Cumulating frequencies shows that 5 670 houses had fewer than 2 bedrooms, $5670 + 15120 = 20790$ had fewer than 3 bedrooms, $20790 + 38430 = 59220$ had fewer than 4, and so on. Or cumulating in the other direction shows that 1 260 houses had 5 or more bedrooms, 3 780 had 4 or more bedrooms, and so on.

We can represent such a distribution pictorially by means of a *line chart*. This is best drawn on graph paper. The values taken by the variable are marked along the horizontal (X) axis as if for a graph. Lines are drawn in a vertical direction (parallel to the Y axis) from this base line, with their lengths proportional to the frequencies $f(X)$. (The diagram will appear the same if the lengths of the lines are drawn proportional to relative frequencies $f(X)/n$.) Figure 2.1 shows the line chart for the distribution of the number of bedrooms (Table 2.1). It is not essential to draw in axes, but is a convenient way of showing the scale of measurement of $f(X)$. This scale *must* be given or the line chart is meaningless.

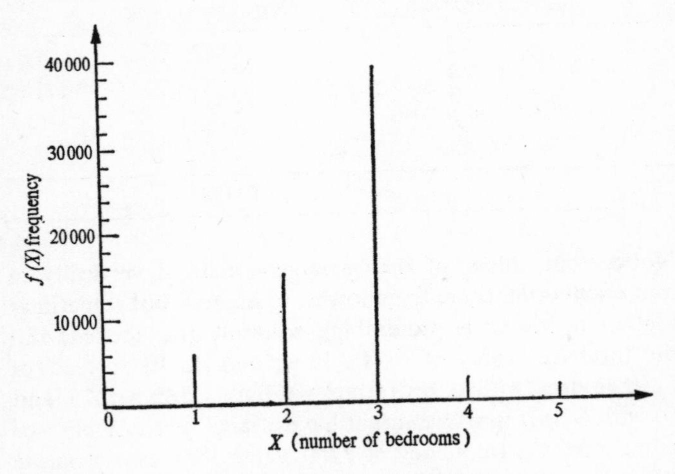

Figure 2.1 (*see Table 2.1*)

Frequency distributions (grouped data)

When the variable is continuous we rarely find that any particular value of the variable occurs more than once, although values may be very close. This is sometimes true of discrete variables too – for example with sums of money ranging from hundreds to thousands of pounds. In these cases we group values into non-overlapping classes of reasonable size, for example, ages of the population of a town might fit well into five-year age groups. We also group discrete variables when there is a large number of values and a large number of observations, for example, the number of words in each article published in a journal during the last ten years.

The size of the classes is determined partly by the nature of the data and partly by the total number of observations. The purpose of grouping is clarification, to bring out broad patterns which may be hidden when individual values are given, but this is at the expense of losing information, since once values are grouped the original values are 'lost' and replaced by classes. It follows that the smaller the number of classes, the greater the loss of information about the values. On the other hand, the larger the number of classes, the less point there is in grouping the values. Generally, somewhere between seven and fifteen classes is sensible, but there is no rule about this. When grouping, the number of classes to use is chosen by a combination of experience and personal judgment.

There are several ways of labelling the classes. An easy way is

0–5
5–10 etc.

Here care must be taken with values which coincide with the end points, that is, the values 0, 5, 10. The usual convention, and one easy to operate, is to put such values in the class above. Thus, values of 5 are put in the class 5–10, values of 10 in the class 10–15, and so on. In fact the labelling means, and could be written as

0 and under 5
5 and under 10 etc.

When the variable is discrete an alternative way of labelling the classes, which avoids ambiguity at the end-points, is

0–4
5–9 etc.

or a similar labelling if the variable takes non-integer values. Here the value 5 clearly goes in the class '5–9', as does also the value 9. This labelling also works for continuous variables, as can be seen by thinking of the discrete measurements of them, for example, length measured to the nearest centimetre. If one of the end-points of a class is not defined the class is said to be *open-ended*, for example, classes 'under 15' and 'over 70' are open ended.

We may on occasions need to represent each class by a single number and we usually do this by representing every value in a class by the mid-point of the class, which we denote by X. The frequency $f(X)$ is then the frequency of the class whose mid-point is X and the relative frequency of the class is $f(X)/n$ where n is the total frequency. To find the mid-point we have to consider all the different values we could put in the class, whether or not they occur in our data. With a discrete variable taking integer values only the class '0–5' (or '0–4') could have in it the values 0, 1, 2, 3, 4 so that its mid-point is the middle value 2. Into the class '0–10' could go values 0, 1, 2, 3, 4, 5, 6, 7, 8, 9 so that its mid-point is $4\frac{1}{2}$, halfway between its two middle values. With a continuous variable the class '0–5' or '0 and under 5' (or a suitable discrete labelling) could have in it any value between 0 and 5 so its mid-point is $2\frac{1}{2}$ (think of dividing 0–5 on a ruler into two equal parts). Similarly the mid-point of '0–10' is 5 when the variable is continuous. When the variable is discrete but each class contains a large number of values so that it becomes impractical to list them all, mid-points are found as if the variable were continuous. For example, money is a discrete variable, but to find mid-points of classes £0–£1 000, £1 000–£2000, . . . representing income groups, we need to treat money as a continuous variable giving mid-points £500, £1 500, ... Diagrammatically classes 0–5, 5–10, have mid-points as shown in Figure 2.2.

Clearly an open-ended class has no mid-point, so that

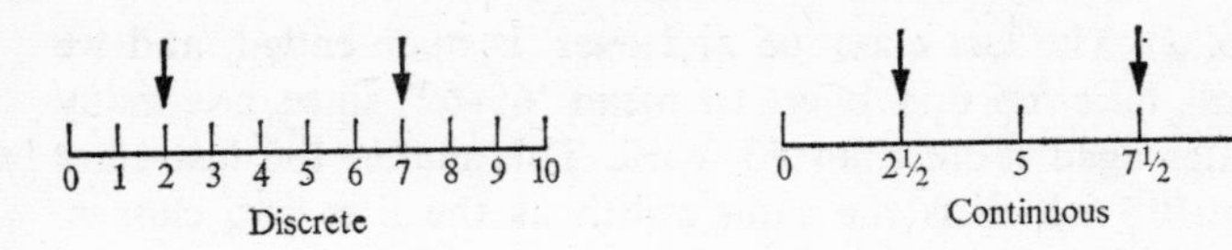

Figure 2.2

when a representative value is needed for the values in an open-ended class it is first necessary to close the class in some sensible manner suggested by the distribution, explaining our reasoning. For example, if the distribution relates to ages of a population it might be sensible to close the class 'over 90' at 105 giving a class '90–105' since very few persons live to be more than 105 and it is fairly safe to assume there is none in any population under consideration.

The *width* of a class is the distance from the beginning of one class to the beginning of the next (in the case of the last class the 'next' class has to be imagined). Again it is helpful to think of a ruler. Just as an open-ended class has no mid-point, it has no finite width either. If the width is needed the class has to be closed as indicated previously.

Since values are replaced by mid-points of classes, ideally classes where all the values are at one end of the class should be avoided. For example, if the majority of values are multiples of 10, classes 10–20, 20–30 and so on, would not be a good idea. Similarly non-occurrence of some values may be important, but could be hidden in some groupings. In practice, as will be seen when operations on frequency distributions are discussed, it is advisable to have classes which are easily defined, as when the end-points are multiples of 5 or 10, and convenient to have classes which are all the same width.

As an example consider the distribution in Table 2.2 which gives the age distribution in years of working males in Great Britain in 1970 (adjusted data from *Social Trends*

No. 2). The last class '65 and over' is open-ended, and we shall take '65 and over' to mean '65–69' since not many males aged more than 70 work. This makes the last class width 5 which is the same width as the first two classes. Notice that the other two classes are width 20, that is, are four times as wide as the first two. (The reader might prefer to argue that say a sizeable number of males between 70 and 75 works and consider the last class to be '65–74'. Any sensible way of closing the class is acceptable.) Table 2.2 also gives the two cumulative distributions.

Table 2.2 Distribution of working males in Great Britain by age

Class (age in years)	*Frequency*	*Cumulative distributions (years)*	
15–19	1 200	$0 < 15$	$0 \geqslant 70$
20–24	1 900	$1\,200 < 20$	$540 \geqslant 65$
25–44	6 340	$3\,100 < 25$	$6\,560 \geqslant 45$
45–64	6 020	$9\,440 < 45$	$12\,900 \geqslant 25$
65 and over*	540	$15\,460 < 65$	$14\,800 \geqslant 20$
	16 000	$16\,000 < 70$	$16\,000 \geqslant 15$

* Closed as 65–69.

Diagrams for grouped frequency distributions

The appropriate pictorial representation of a grouped frequency distribution such as that in Table 2.2 is a *histogram*. The classes are marked along the horizontal axis. Every class is taken to extend to the beginning of the next class when the diagram is drawn, whatever the labelling, so that only the beginnings of classes need be indicated. Rectangles are then constructed on the classes marked so that their *areas* are proportional to the class frequencies. Notice that classes are marked so that the bases of the rectangles are proportional to class widths. When all the classes are of the same width drawing rectangles of heights proportional to frequencies automatically gives rectangles with areas proportional to frequencies. When the classes are not all the same width an adjustment has to be made in the height. For example, if one class is three times as wide as all the others then, for a given frequency, the rectangle for that class needs to be a third of the height to ensure that its area is of the correct size. If there are open-ended classes in the distribution these have to be closed before a histogram can be drawn.

The histogram of the distribution given in Table 2.2 is shown in Figure 2.3. It is customary to indicate frequencies along the *Y* axis. We see that the frequencies on the vertical axis imply a frequency of 1 585 in each of the classes 25–29, 30–34, 35–39, and 40–44. Combining these gives a frequency of 6 340 in the class 25–44 as desired. Similarly the frequency of 6 020 in the class 45–64 appears as a frequency of 1 505 in each of four classes of width 5. An alternative to showing frequencies on a vertical axis is to write the frequency of each class at the top of the appropriate rectangle. Frequencies must be indicated in some way.

The classes must also be labelled. An alternative to labelling the end points is to write the class name under the appropriate rectangle. This is especially useful with a distribution like this given in terms of classes of the type 0–4, 4–9, ... (which are drawn as if they had the labelling 0–5, 5–10, ...). The only way in which there can be a gap between rectangles is when the gap corresponds to a class of zero frequency.

The histogram will appear the same if areas are drawn proportional to relative frequencies. Adding up all the relative frequencies gives a total of 1, so we can think of the area of each rectangle as representing the relative frequency of a class, and the total area of the histogram as representing 1. If the area is thought of in this manner the link between

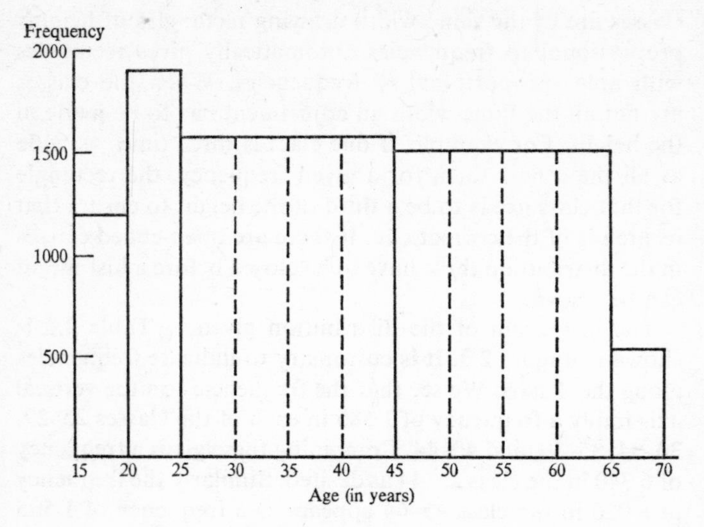

Figure 2.3 (see Table 2.2)

histograms and probability distributions is easier to understand (see chapter 5).

The *frequency polygon* is another diagrammatic representation of a grouped frequency distribution. A quick way of describing it for distributions where the classes are all the same width is to say that it is obtained by joining the midpoints of the tops of the rectangles in a histogram by straight lines as shown in Figure 2.4. Part of each rectangle is excluded but for each triangle cut off a triangle of equal area is added. The diagram shows how to extend the polygon to the horizontal axis so that triangles of the correct size are added. When classes are not all the same width an adjustment has to be made in the position of the points plotted to ensure that the area under the polygon is the same as that under the histogram. There is no need to draw a histogram in order to draw a frequency polygon since the positions of the points are easily found. In fact, an alternative way to describe a frequency polygon is to say that its points have as abscissae (X values) the mid-points of the classes, and as ordinates (Y values) the frequencies or relative frequencies of the classes. It is easier to compare two or more different frequency polygons drawn on the same diagram than two or more histograms.

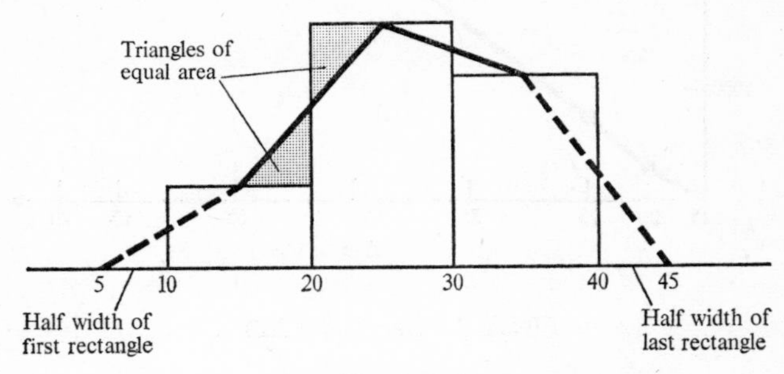

Figure 2.4

A distribution can be plotted in terms of cumulative frequencies and the resulting diagram is called a *cumulative frequency curve* or *ogive*. The ogive, qualified as 'less than' or 'more than or equal to' is best plotted on graph paper. Its points have end-points of classes as abscissae and cumulative frequencies as ordinates. Thus the point (25, 3 100) is a point on the 'less than' ogive drawn in Figure 2.5 for the cumulative distribution of Table 2.2. Notice that the points are joined by straight lines. The axes must be labelled and their scales marked.

Tabulation

The frequency distributions discussed so far, including frequency distributions constructed from qualitative data,

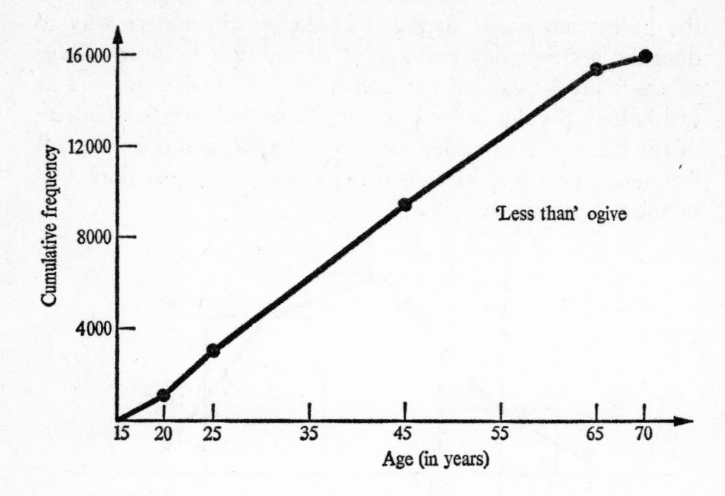

Figure 2.5 (*see Table 2.2*)

can all be described as one-way tables. We can, of course, classify data in more than one way. When we sort data by two factors of classification, for example, count how many manual workers are in each of a set of specified income groups, and how many non-manual workers are in each income group, the resulting table of frequencies is said to be two-way. Data sorted by three factors of classification give rise to a three-way table, and so on. Qualitative and quantitative variables can both appear in the same table (see question 3). The Census of Population provides many examples of tables which have several factors of classification and which occupy numerous pages.

A table which is well set out helps clarify information in the data. One or two minor points to notice which, if followed, will ensure that the table is well presented, are as follows. All rows and columns should be headed and the headings must be either self-explanatory or explained in footnotes. Units of measurement should be clear. Tables require titles, and where appropriate the source of the table should be given. When it is sensible to find totals or subtotals these should be included in the table. Tables that repeat headings unnecessarily should be avoided. There are generally several satisfactory ways of setting out tables, but sometimes a change in lay-out will emphasise a different point, or enable a different comparison to be made more easily. Table 2.3 is an example of a two-way table.

Table 2.3 Live births registered 1970 (in thousands)

	England and Wales	*Scotland*	*N. Ireland*	*UK*
Males	403	45·0	16·5	464·5
Females	381	42·3	15·6	438·9
Total	784	87·3	32·1	903·4

Source: Central Statistical Office, *Facts in Focus*, 1972.

Tables are often presented in terms of proportions (relative frequencies), normally in the form of percentages as these are readily understandable and enable quick comparisons to be made. Percentages can be formed by converting frequencies to percentages of row totals, or to percentages of column totals, or to percentages of the overall total. The different percentages serve different purposes. For example, referring to Table 2.3 (all figures in thousands), 403 live births registered in 1970 were male births in England and Wales. In all there were 903·4 births, so 44·6% of all births in the United Kingdom were male births and were in England and Wales. Of the 784 births in England and Wales 51·4% were male births, and of the 464·5 male births in the United Kingdom, 86·8% occurred in England and Wales. It is necessary to make clear what is a percentage of what and relevant totals should be given so that frequencies can be derived from percentages if required. For example, knowing that 51·4% of 784 births were males enables us to calculate how many were male births, but simply knowing that 51·4% of births were males does not (for this might mean that 5 140 out of 10 000 births were male, for instance). Thus a percentage in itself can be misleading.

Diagrams for qualitative distributions

There are three main ways of representing qualitative distributions by diagrams. The most satisfactory way is by the *bar chart*. Parallel bars of equal width are drawn with lengths proportional to frequencies, one bar for each subdivision of the variable. The bars are normally drawn vertically or horizontally on the page, but to avoid confusion with histograms it is suggested that bars in bar charts are always drawn horizontally. The bars may touch one another, but again it avoids confusion to draw them separated and is neater. The bars should be labelled and the frequencies given, either on the bars or in the form of a distribution. If preferred, the bars may be considered to be proportional to relative frequencies which gives exactly the same picture. Bar charts are most easily constructed on graph paper.

When representing by a diagram a two-way table where both factors of classification are qualitative there is some scope for ingenuity. Suppose, for example, that we want to represent Table 2.3 diagrammatically. We shall combine the figures for Scotland and Northern Ireland since these are very small compared with England and Wales. Here we could, for instance, draw one set of bars for male births and one set for female, or could draw a set of bars for areas and subdivide each bar to show how figures for male and female births were combined to give the total. Figure 2.6 illustrates two ways of drawing bar charts in this situation. Other ways can be devised. The first shown enables easy comparison of the number of male births with the number of female births in any one area, and easy comparison of the number of males (or females) born in any area with the number born in any other area. The second bar chart drawn enables easy comparison of one area with another both as regards the number of male births and the total births, but the comparison for female births is less easy. The form of bar chart to draw depends on which comparisons are of most interest. When more than two factors of classification are involved a larger selection of bar charts on lines similar to these can be drawn.

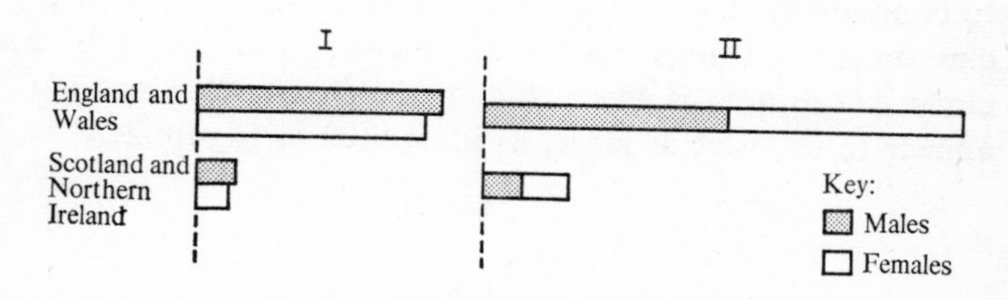

Figure 2.6 (*see Table 2.3*)

Similar to the bar chart is the *pictogram*. In this a picture or symbol represents a specified number of items instead of a unit of length as in the bar chart, so that instead of a bar we draw a line of symbols. Pictograms are supposed to have popular appeal, but, apart from being unsuitable for manual drawing, have a number of points against them. If more than one kind of symbol is to be used in a pictogram, these should be equivalent in size and to some extent shape, or the diagram will distort the data. For example, a pin woman wearing a skirt will appear to represent a greater number than a pin man with less bulky clothing. If a fraction of a symbol is needed this will be difficult to represent accurately. Spacing between different types of symbols must be equivalent, or the relative lengths of the lines of symbols will be wrong. The bar chart does well what the pictogram does badly and is to be preferred.

The third way of representing qualitative frequency distributions diagrammatically is by the *pie chart*. This is a circle divided into segments of areas proportional to frequencies. It is easy to understand, and accurate, but the calculations involved in order to draw it (frequency f is represented by a segment whose angle at the centre is $(f/n) \times 360°$ where n is the total frequency) can be messy. The total area of the circle is proportional to the total frequency. Another possible objection to pie charts is that if we want to compare circles representing two different total frequencies optical illusions distort the picture. For example, a circle whose area is twice that of the first circle does not appear to be twice as large, as illustrated in Figure 2.7.

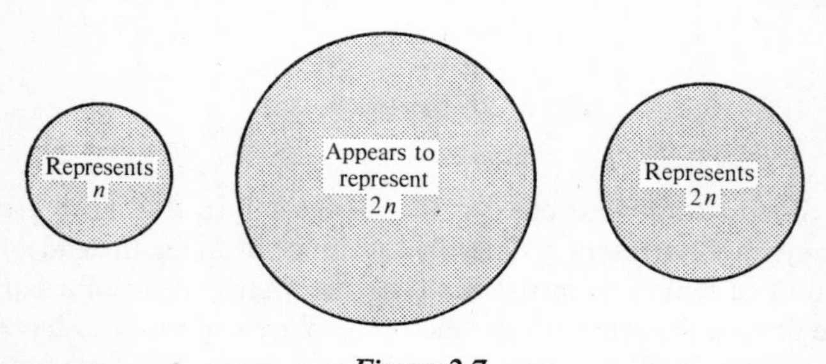

Figure 2.7

Time Series

When we have a series of observations over time, for example, road mileages for each day, unemployment figures for each month, sales figures for each quarter, population sizes for each year, these can be plotted as a *time series* as in Figure 2.8 which relates to the number of houses demolished in clearance areas and unfit houses demolished or closed elsewhere (in thousands) in England and Wales during different years (data given in Central Statistical Office, 1972 and reproduced with diagram). Time is plotted on the *X* axis, and the values on the *Y* axis. The points are joined by straight lines to indicate continuity over time. As with all graphs the axes must be labelled and their scales marked. Note that while time itself is a continuous variable, in practice our observations may refer to intervals. Thus National Income for 1972 refers to the income generated throughout that year. On the other hand the population of a country in 1972 is the population at a particular point in time in 1972.

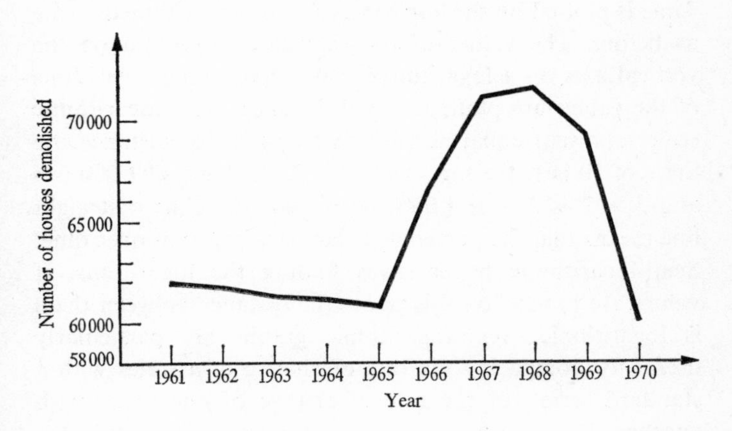

Number of houses demolished in clearance areas and unfit houses demolished or closed elsewhere (in thousands) in England and Wales

Year	1961	1962	1963	1964	1965	1966	1967	1968	1969	1970
No. houses	62·0	61·8	61·4	61·2	60·7	66·8	71·2	71·6	69·2	60·2

Figure 2.8

If we think of a particular example, such as unemployment, it is clear that different components are concealed in the variation of the observations O over time. There is possibly a long-term trend T, possibly a seasonal component S, perhaps a cyclical component C, and an irregular (or error) component I. Study of the time series diagram enables us to make sensible comments about these components. Analysis of time series is outside the scope of this book, but is covered in for example, Hoel (1971), Wonnacott and Wonnacott (1972), and Yeomans (1968).

When we are interested in percentage changes of a variable over time, or equivalently, ratio changes[1] the appropriate way of representing the series is on a *semi-logarithmic graph*. Time is plotted on the horizontal axis on an arithmetic scale as before. The values of the variable are plotted on the vertical axis on a logarithmic scale, that is, the logarithms of the values are plotted. Equal distances on a logarithmic scale represent equal percentage changes (for instance the $\log_{10}$ of 10 is 1, the $\log_{10}$ of 100 is 2, the $\log_{10}$ of 1 000 is 3 and $3 - 2 = 2 - 1$, $1\,000/100 = 100/10$). Thus a straight line means that the percentage change is constant over time. Semi-logarithmic paper saves finding the logarithms. If values are plotted on this paper the distance between them is logarithmic. Semi-logarithmic graphs are particularly useful for comparing the rate of change of a series with a standard series, or the rate of change of one series with another.

[1] If $X(t)$ and $X(t+1)$ are the values of the variable at successive times t and $t+1$ respectively, the percentage change from t to $t+1$ is $\frac{X(t+1)-X(t)}{X(t)} \times 100$ and the ratio change is $X(t+1)/X(t)$. Percentage change is 100 (ratio change minus one). If the one measure remains constant over time, so must the other.

Exercises

1 For each of the following say

a on what scale of measurement it is measured

b where appropriate, whether it is continuous or discrete.

i the number of employees in a firm
ii the proportion of the population unemployed
iii class of degree
iv the number of rooms in a house
v the rank of an army officer
vi the type of tenure of a dwelling
vii weekly expenditure on food
viii the position of a trade union (ordered by size)
ix weights of a commodity
x the location of branches of a large retail concern
xi nationality
xii measurement of intelligence

2 A travel agent calculates, from his records for the summer months of May to September inclusive, average numbers of travellers departing, as follows. On week-days (average over Mondays to Thursdays) the average number of persons departing by air to Europe was 30, and by ship to Europe was 21, whereas the average number going elsewhere was 15 both for air departures and for ship departures. On Fridays the average number going to Europe by air was 75, on Saturdays it was 60, and on Sundays it was 42. The corresponding figures for air departures to other destinations were 10, 9, and 7. On Fridays the average number of departures by ship to Europe was 39 and by ship elsewhere 20. The corresponding Saturday figures were 39 and 33, and the Sunday figures were 18 and 29. Put this information in the form of a table.

3 The table below gives the size of manufacturing establishments by numbers of employees for four different types of

industry. Each number can be found (*a*) as a percentage of its row total, (*b*) as a percentage of its column total, (*c*) as a percentage of the overall total. Interpret these different percentages and calculate some of them in order to comment on the main points shown by the table. (Every percentage works out exactly to the second decimal place.)

	Size of establishment (no. of employees)				
Type of industry	0–49	50–99	100–499	500–999	1 000 *and over*
Food, drink and tobacco	1 976	1 716	1 284	156	104
Vehicles	684	726	408	114	68
Textiles	1 862	1 386	1 848	294	210
Clothing and footwear	3 078	2 772	1 296	36	18

4 Referring to published tables, investigate migration between the different regions of Great Britain and the factors affecting migration.

5 In a small survey of 50 families with children of school age the number of children under fifteen years old in each family was as follows:

2	2	5	4	2	4	3	3	4	7
2	2	2	5	8	4	1	2	3	3
2	7	2	5	3	1	4	2	5	4
5	2	4	3	5	5	3	9	5	2
1	1	6	2	3	2	5	2	2	3

Form a frequency distribution from these data and present it in diagram form also. In addition, present the distribution in terms of relative frequencies and in terms of cumulative frequencies.

6 A small firm is considering improving its catering facilities and wishes to know how many employees use the canteen at present. On 70 successive days the number of employees eating in the canteen was recorded to be (read left to right, row by row):

58	57	61	57	58	62	57	58	58	57
58	62	60	63	57	59	57	58	58	58
63	61	64	58	63	61	63	63	58	64
58	61	58	61	63	59	62	57	60	63
59	57	64	59	57	63	62	64	57	58
59	58	57	64	57	57	59	59	60	63
61	63	62	64	60	59	62	62	62	62

Form a frequency distribution from these data and present it in diagram form also. In addition present the distribution in terms of relative frequencies and in terms of cumulative frequencies.

7 The numbers below are the numbers of questions answered correctly by each of 60 persons who took part in a quiz of 99 questions.

57	59	41	50	23	97	16	35	58	48
48	26	75	46	32	53	52	67	60	61
42	38	12	72	59	68	27	68	27	74
85	43	61	58	70	71	62	79	40	63
7	50	58	73	92	44	84	90	83	73
43	68	24	89	55	35	66	45	72	56

Set up a frequency distribution using 10 equal-sized intervals. Give the mid-points of the classes and their common width. Draw the histogram and the frequency polygon.

Construct a cumulative frequency distribution and draw the ogive.

8 The distribution below shows, for men, the percentage in different income groups. Give the mid-points and widths of the classes. Give a cumulative distribution. Draw the histogram and appropriate ogive.

Income (in £ per week)	*% of men with income shown*
Under 7	0·5
7–11	17·6
11–15	39·9
15–19	26·4
19–23	10·8
23–30	4·2
30–40	0·4
40 and over	0·2
	100·0

9 Below, is the age distribution, in the form of proportions, of females employed in retail distribution. Give the mid-points and widths of the classes. Draw the frequency polygon and an ogive.

Age (in years)	*Relative frequency*
15–20	0·21
20–25	0·16
25–30	0·08
30–35	0·08
35–40	0·10
40–45	0·12
45–50	0·10
50–55	0·07
55–60	0·05
60–65	0·03
	1·00

10 Represent the following diagrammatically:

a *Animals slaughtered in a year (thousands)*

Cattle and calves	42
Sheep and lambs	132
Pigs	107

b *Sunshine in mean hours per day*

Year	1	2	3	4	5	6	7	8	9
Sunshine	4·20	4·16	4·25	4·25	3·95	3·94	4·16	4·20	4·30

Year	10	11	12	13	14	15	16	17	18
Sunshine	4·14	4·20	3·81	4·02	4·20	3·74	3·62	3·48	4·06

c *Industrial stoppages in a year*

Duration of stoppage	*Number of workers*
Not more than 6 days	352
Over 6 but not more than 12	94
Over 12 but not more than 24	23
Over 24 but not more than 36	9
Over 36 but not more than 60	2
Over 60 days	0

d *Number of men and women students at a provincial university classified by type of residence*

	Men	*Women*	*All students*
Halls	1 069	394	1 463
Other university accommodation	318	130	448
Lodgings	1 301	597	1 898
Home	178	58	236
All types of residence	2 866	1 179	4 045

e *Sales and profit data for a firm* (*the diagram should show rates of change*).

Year	1	2	3	4	5
Sales (1 000s)	150	302	490	199	256
Profit (£100s)	58	150	207	101	120

3

Descriptive statistics
Part 2 — Measures of location and dispersion

When we begin to examine frequency distributions of quantitative variables and the diagrammatic representation of them, and particularly when we begin to compare different distributions, a natural question to ask is whether we can find two or three numbers which in some way summarize each distribution. In one distribution the values might all be relatively high, in another all relatively low, so that one desirable summary measure would be a figure which is in some way an average and which indicates the position or location of the values on a numerical scale. In one distribution the values might all be close to one another numerically, but in another they might be spread over a large numerical interval. Thus a summary measure which indicates the spread, or dispersion, or variability of the values in the distribution would also be useful, and possibly this measure should consider the spread about an average value.

We might also like a measure telling us about the shape of the distribution. Is the distribution symmetrical or is it skew? If a distribution is *symmetrical* we can draw a line parallel to the frequency (Y) axis on the line chart, or histogram, so that the picture on one side of the line is a reflection of that on the other side. If a distribution is not symmetrical it is said to be *skew*, that is it is 'pushed off centre'. When the higher frequencies occur at the lower values the distribution is said to be skewed to the right or to have a positive skew. When the higher frequencies occur at the higher values the distribution is said to be skewed to the left or to have a negative skew. Figure 3.1 illustrates with histograms four basic shapes of distribution. (See Yeomans, 1968, for measures of skewness.)

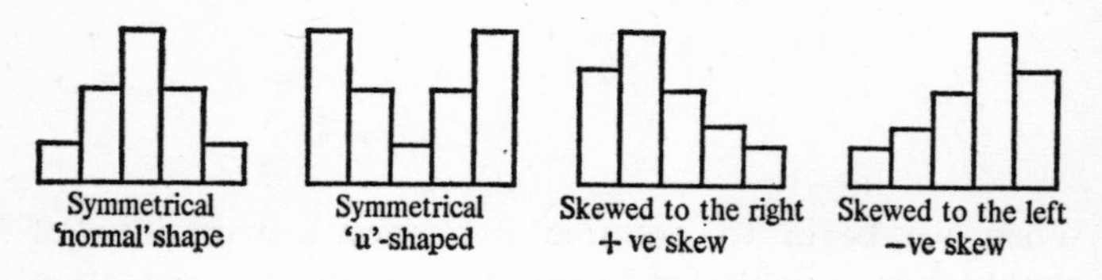

Figure 3.1

The mode

The first measure of position or average measure we consider is the *mode*. With qualitative data the mode is the identifier of greatest frequency, with discrete data that are ungrouped the mode is that value X for which the frequency $f(X)$ is highest. Thus, in a sense, the mode is the most fashionable value ('*à la mode*' means 'in the fashion'). For example, in the distribution of number of bedrooms (Table 2.1, p. 13) the mode is 3, that is, more houses have three bedrooms than any other number of bedrooms. On the line chart the mode is the value corresponding to the longest line.

With grouped data the *modal class* is usually defined to be the class of highest frequency, but this can be unsatisfactory as a definition if the classes are of very different widths, for we expect the frequency to be higher in a wider class just because there are more values which could fall in that class. Thus, it might be better to adjust the distribution into one having classes which are all of the same width (as when drawing a histogram) and take the class of highest frequency in the adjusted distribution as the modal class. This means that open-ended classes need to be closed, that is, given widths when determining which class is the modal class. In the distribution of working males in Great Britain by age (Table 2.2, p. 18) the class of greatest frequency is 25–44, but this class is one of the two wider classes. Adjusting the distribution to make all classes of width 5 gives the class 20–24 as the modal class.

If we require a single value for a mode from a grouped distribution we can take the mid-point of the modal class as that value. However, we might notice that of the two classes adjacent to the modal class one is of higher frequency than the other and it seems reasonable to suppose that the mode will lie nearer to the adjacent class of higher frequency than to the other class. Thus, with the data on working males by age, we expect the modal age to be nearer 25 than 20.

The diagrammatic method of finding a single value for the mode which takes account of the frequencies of the two classes adjacent to the modal class is shown in Figure 3.2. Provided the diagram is neat and of a reasonable size the value found in this way will be accurate enough for most purposes. The modal class must be labelled so that the mid-point of the class falls at the mid-point of the interval representing it. This will ensure that when the two adjacent classes are of equal frequencies the value of the mode obtained by this method is the mid-point of the modal class in accordance with common sense. Notice that with discrete data the mode is sometimes a value that cannot occur. A formula for calculating the value of the mode can be found by considering the geometry of the diagram (see Yeomans, 1968).

The mode is an easy measure to understand and it is quick to find. It is the only average measure that can be

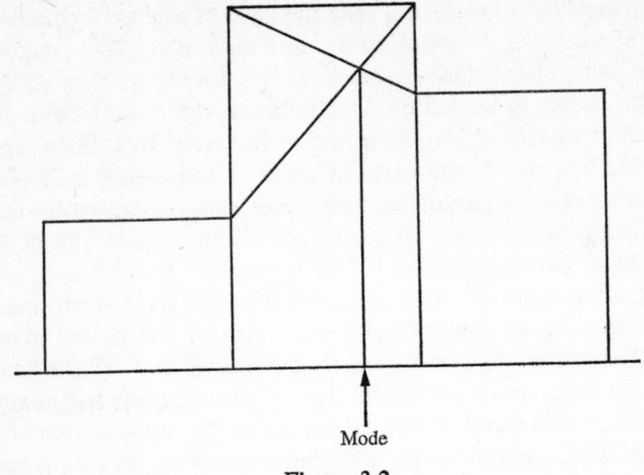

Figure 3.2

found for qualitative distributions. When statements such as 'the average man watches football on Saturdays' are made, the average referred to is the mode. Some distributions have two or more modes. A distribution with two modes, such as a U-shaped distribution, is said to be bimodal. The distribution of ages of pedestrians involved in road accidents is an example of a U-shaped distribution. Most accidents occur to the very young and to the very old. When dress or shoe sizes are described as average, this means that most people need dresses or shoes of that size – another use of the mode.

The chief points against the mode as an average measure are that it is not very stable in the sense that its value can be changed fairly considerably by changing the class end-points and when classes are not all the same width there are difficulties in finding it. The more skewed the distribution, the worse is the mode as a measure.

The range

The simplest measure of dispersion or spread is the *range*, and this is the appropriate measure of spread to use with the mode. The range is the difference between the lowest and highest values, either given as L to H, where L is the lowest value and H is the highest, or as the numerical difference H minus L. The first presentation clearly gives more information and the subtraction can then be done if a single number is desired. The range of the distribution of number of bedrooms (Table 2.1, p. 13) is 1 to 5, that is, the range is 4 (for $5-1 = 4$).

The range is not a satisfactory measure for grouped distributions where the exact values are not known. For such distributions we have to take as L the lowest value which could occur in the first class of non-zero frequency, and as H the highest value which could occur in the last class of non-zero frequency, and we might have to close classes in order to decide on the lowest and highest values which could occur. For example, in the distribution of ages of working males in Great Britain (Table 2.2, p. 18) the first class of non-zero frequency is the class '15–19' and the lowest value which could occur is 15, so we take L to be 15. The last class of non-zero frequency is the class '65 and over'. If we take '65 and over' as meaning '65–69' the highest value that could occur is 69 so we take H to be 69, but if we take '65 and over' as meaning '65–74' say, the highest value that could occur is 74, so we take H to be 74. Thus the range of the distribution is 15 to 69 with the first closing of the class '65 and over', and 15 to 74 with the second.

When actual values of L and H are known and given with a grouped distribution the range can be found exactly, and in addition we have a much better idea of the distribution of values. For example, with the distribution under discussion, if we are told that the ages ranged from 16 to 67 years, a range of 51 years, we know that the class '65 and over' is in fact the class '65–67'.

Like the mode, the range is easy to understand and quick to find (but it does not exist for qualitative data). The mode and range together give a quick summary of a distribution and finding them is a useful first step in analysing a set of figures, especially in surveys. The chief disadvantage of the range is that it is greatly influenced by atypical extreme values. For example, with the distribution of bedrooms, if there were another house with 25 bedrooms, the range would be 1 to 25 which would suggest there were houses with number of bedrooms somewhere between 5 and 25 – far from the actual situation. In such a case it would be better to say that the number of bedrooms ranges from 1 to 5 and one house has 25 bedrooms.

The median

The second average measure we consider is the *median*. This might be described as a 'half-way' measure, for it is a value such that half the values are less than it (and half the values are greater than it). If we have a set of discrete values arranged in order of magnitude the median is the middle value listed when there is an odd number of values, for example, in the five values 1,1,2,3,5, the median is 2. When there is an even number of values the median could be any value between the two middle values listed, but by convention it is generally taken to be the value that is halfway between the two middle values, for example, in the six values 1,1,2,3,3,5, the two middle values are 2 and 3 (the first figure 3 listed) and the median is generally taken to be $2\frac{1}{2}$ (though $2\frac{1}{4}$ or $2\frac{3}{4}$ say would satisfy the definition just as well).

When a set of discrete values is arranged in a frequency distribution there is no need to ungroup the values and list them separately in order to find the median. We can easily find the position of the median. If the total frequency is odd then it can be written as $2m+1$, where m is an integer, and the median is the $(m+1)$th value when the values are listed in order of magnitude. If the total frequency is even then it can be written as $2m$, and the median is halfway (by convention) between the mth value and the $(m+1)$th value when the values are listed in order of magnitude. For example, if the total frequency is 51 (odd), $51 = 2 \times 25 + 1$ and the median is the 26th value. If the total frequency is 92 (even), $92 = 2 \times 46$ and the median is halfway between the 46th value and the 47th value.

The cumulative frequency distribution tells us what values are taken in the mth and $(m+1)$th positions. Refer to the distribution of bedrooms (Table 2.1, p. 13). The total frequency is $63000 = 2 \times 31500$ so for the median we want the value halfway between the 31500th and 31501th values. Cumulating frequencies shows that 20790 houses had fewer than 3 bedrooms, and 59220 had fewer than 4 bedrooms. Thus both the required values are 4 and the median number of bedrooms is 4. Notice that cumulating frequencies is another way of saying that the first 5 670 values are 1, the 5670th to 20790th values are 2, and so on.

The procedure is similar with values arranged in a grouped distribution when we have access to the ungrouped data. We first calculate the position of the median, and then use the cumulative distribution to find the class in which the median lies. For example, suppose that the total frequency is 51 so that the median is the 26th value. Suppose that the cumulative frequency distribution tells us that 24 values are less than 38, and 30 values are less than 40, so that the 26th value is in the class 38–40. If the $30–24 = 6$ values in the class 38–40 are, in order of magnitude: 38·0, 38·2, 38·7, 39·1, 39·3, 39·8: then the 25th value is 38·0 and the 26th value is 38·2. Thus the median (26th value) is 38·2.

To find the median of grouped distributions when the

ungrouped values are not available, the first step in calculating the value of the median is again to find the class in which the median lies, but the procedure is now slightly different. Since the median is a value such that half the values are less than it, if we consider a histogram where area represents frequency, the line parallel to the frequency axis dividing the area of the histogram in two must mark the position of the median whether the total frequency is even or odd. For example, if the total frequency is 51 we must find the value with frequency $25\frac{1}{2}$ less than it. If the total frequency is 92 we must find the value with frequency 46 less than it. We do not find this value by examining the histogram, however, but by calculation as described below.

Suppose that the total frequency is n, then the median is the value with frequency $n/2$ less than it. As before, cumulating frequencies indicates in which class the median lies. Suppose that it lies in a class which begins at the value L and that n' values are less than L. Suppose, further, that the class is of width c and of frequency f. (If by any chance the class is open-ended it will be necessary to close it at this stage.) We assume that the f values in the class are evenly distributed, so we imagine that the block of the histogram is divided into f strips parallel to the frequency axis, each strip being of width c/f. We need to go $(n/2-n')$ frequencies into the class which is a distance $(n/2-n')c/f$. Thus the median is the value $L+(n/2-n')c/f$. The reader is advised to remember how to find the median rather than learn the formula.[1]

[1] The reader who is puzzled by the apparent contradiction between this and the procedure for ungrouped data should notice that each value is assumed to occur at the mid-point of a strip. When n is odd, e.g. $n = 51$, we need to go (a whole number$+\frac{1}{2}$) frequencies into the class which brings us to the mid-point of the strip representing the 26th value. When n is even, e.g. $n = 92$, we need to go a whole number of frequencies into the class which brings us to the point halfway between the mid-points of the strips representing the 46th and 47th values.

Referring to the distribution of the ages of working males (Table 2.2, p. 18) the total frequency is 16000 so the median is the value with frequency 8000 less than it. Cumulating frequencies shows that 3100 males are less than 25 and 9440 males are less than 45. Therefore the median lies in the class beginning at 25. The frequency of this class is 6 340 and its width is 20. We need to go 8000 − 3100 = 4 900 frequencies into the class, or a distance 4 900 times (20/6 340) = 15·46 years (rounded to second decimal place). Therefore the median age is 25 + 15·46 = 40·46 years.

The median can be found diagrammatically from the ogive, and provided the diagram is neat and not over small, the value found will not differ greatly from the calculated value. The method is illustrated in Figure 3.3.

Similar to the median are quartiles, deciles, and percent-

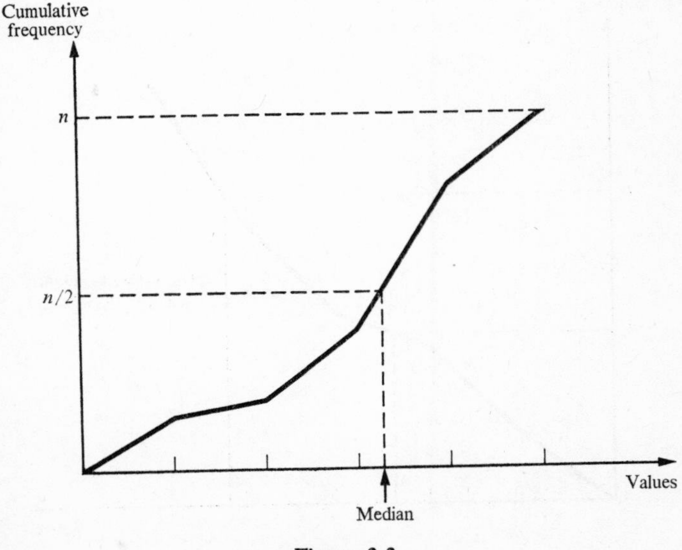

Figure 3.3

iles. There are three quartiles and they are values dividing the frequency into four equal parts (just as the median divides the frequency into two equal parts). Deciles divide the frequency into ten equal parts, and percentiles divide the frequency into one hundred equal parts. The median, quartiles, deciles, and percentiles are all particular kinds of *quantiles.* All can be calculated or found diagrammatically using the same method as described for the median replacing $n/2$ by some other fraction of n, for example $n/10$, $2n/10$, $3n/10, \ldots$ if the deciles are required.

The ogive can also be used to find frequencies from values. As an example Figure 3.4 illustrates how to estimate the number of values between values x_1 and x_2. This is the (cumulative) number less than x_2 minus the (cumulative) number less than x_1.

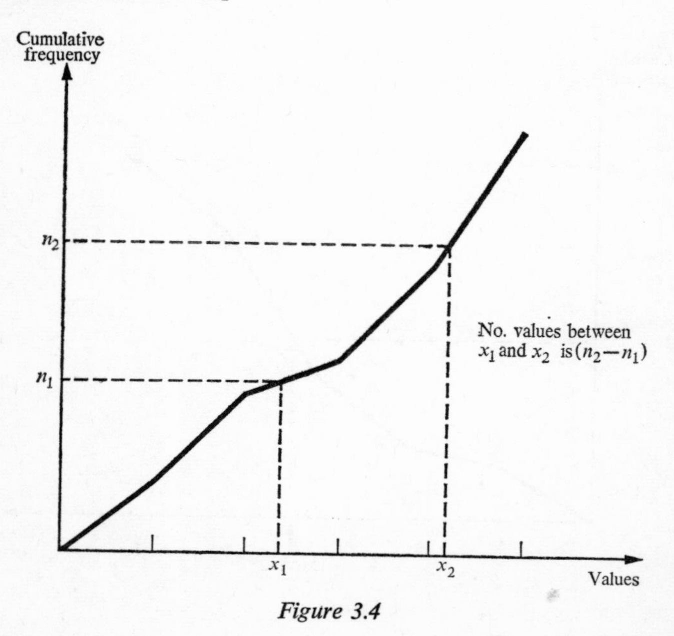

Figure 3.4

The concept of the median is easy to understand and the median, and possibly quartiles, are often used naturally in describing situations. For example, we might notice that in an examination although half the candidates obtained a mark less than 40 per cent, a quarter of the candidates obtained a mark greater than 70 per cent. Quantiles are used a great deal in connection with income distributions, which are typically skewed to the right. The median household income is such that half of all households receive less than it, and the first quartile income is such that a quarter of all households receive less than it. If we describe these last households as 'poor', comparing the first quartile with the median shows how the poor are in comparison with households on average. Similarly, the median is a useful average in other social welfare applications. It is the appropriate one to use when we require averages such as the average number of children in classes, or the average number of floors in buildings. Experimental scientists refer to the median lethal dose (MLD) which is the dose needed to kill half the animals under study.

Change of grouping does not change the value of the median very much. It is unusual for a median to fall in the first or last class of a distribution, so that if these classes are open-ended there is normally no need to close them in order to find the median. Similarly, the value of the median is not influenced by atypical extreme values. It is the frequencies only, not the values, at the lower (or higher) ends of the distribution which determine the median. In addition, the median is fairly easy and quick to calculate, but unfortunately it is not very easy to use algebraically.

One further fact relating to the median is described below. Suppose we have a set of values X and the median is M. Suppose we find the distances of the values X from the median. When X is bigger than M the distance is $X-M$, when X is equal to M the distance is zero, and when X is smaller than M the distance is $M-X$. We can describe the distance of X from M mathematically as the modulus of $(X-M)$ which we write $|X-M|$. Consider any other value M' and find the distances of the values X from M'. The sum of the distances of values X from M' will be larger than the sum of the distances of values X from M whatever the value of M'. This minimum distance property of the median is sometimes useful.

The semi-interquartile range

The appropriate measure of spread to use with the median is the *semi-interquartile range*, which has the alternative name of *quartile deviation*. The three quartiles can be denoted by Q_1, Q_2, and Q_3. The first quartile, Q_1, is the value such that a quarter of the values are less than it. The second quartile, Q_2, is the value such that two quarters, that is one half, of the values are less than it. Thus Q_2 is the median. The semi-interquartile range is defined to be half the distance between the first and third quartiles, that is, it is $\frac{1}{2}(Q_3-Q_1)$. It is in fact half the sum of the distance of Q_1 from Q_2 i.e., (Q_2-Q_1) and the distance of Q_3 from Q_2 i.e., (Q_3-Q_2).

Refer again to the age distribution of working males (Table 2.2, p. 18), for which we have already found Q_2 (median) to be 40·46 years. Since the total frequency is 16 000, Q_1 is the value with frequency $\frac{1}{4}\times 16\,000 = 4\,000$ less than it, and Q_3 is the value with frequency 12 000 less than it. Using a similar method to that used to calculate the median we find that

$$Q_1 = 25+(4\,000-3\,100)\left(\frac{20}{6\,340}\right)$$

$$= 27{\cdot}839 \text{ years (unrounded)}$$

$$Q_3 = 45+(12\,000-9440)\left(\frac{20}{6\,020}\right)$$
$$= 53{\cdot}504 \text{ years (unrounded)}$$

$\therefore$ Semi-interquartile range $= \frac{1}{2}(Q_3-Q_1)$
$= \frac{1}{2}(53{\cdot}504-27{\cdot}839)$
$= 12{\cdot}83$ years (correct to 2nd decimal place)

The arithmetic mean

The third measure of location we describe is the *arithmetic mean.* The layman commonly refers to this as the 'average', although in statistical language the mode and the median are also averages. Sometimes the arithmetic mean is called simply the 'mean', but must not be confused with the geometric mean or harmonic mean (see Bugg *et al.*, 1968).

The measure known to the layman as the average, and probably familiar to the reader as, for example, the average age of a class of school-children, or an average examination mark, and known to the statistician as the arithmetic mean, is the measure found by adding together all the values and dividing by the total number of values. If the values are denoted by X we denote the arithmetic mean by $\bar{X}$ (pronounced X bar) and

$$\bar{X} = \frac{\Sigma X}{n}$$

where n is the total frequency (see Appendix 2 for explanation of Σ notation). If the value X occurs with frequency $f(X)$ clearly

$$\bar{X} = \frac{\Sigma Xf(X)}{n} = \frac{\Sigma Xf(X)}{\Sigma f(X)}$$

Notice also that

$$\bar{X} = \Sigma X \frac{f(X)}{n} = \Sigma X \text{ times relative frequency}$$

The arithmetic mean is a 'sharing out average'. If all the values are pooled so that the total amount available is $\Sigma Xf(X)$, and then shared out equally, each of the n units receives amount $(\Sigma Xf(X))/n$. We might say this is the amount each unit expects to receive. Alternatively, we can think of the arithmetic mean as being the centre of gravity of the distribution.

Notice that if every value is made smaller by an amount c (c is a fixed constant) we have a set of values $(X-c)$ and the arithmetic mean of the new values is smaller than $\bar{X}$ by amount c. Thus to get $\bar{X}$ we must add c to the mean of the new values. This fact sometimes makes arithmetic easier. For example, suppose we want the arithmetic mean of the numbers 93, 98, 101, and 104. Subtracting 90 from each value gives as new values the numbers 3, 8, 11, and 14. The arithmetic mean of the new values is

$$\tfrac{1}{4}(3+8+11+14) = \tfrac{1}{4}(36) = 9$$

Thus the mean of the original values is $90+9 = 99$. The change from values X to values $(X-c)$ is called 'changing the origin'. Values X are measured as distances (with direction indicated as positive or negative) from the origin of zero. Values $(X-c)$ are measured as distances from the origin of c.

If we divide every value X by a fixed constant k, giving a new set of values X/k, the arithmetic mean of the new values is $\bar{X}/k$, so to get $\bar{X}$ we must multiply the mean of the new values by k. This change from values X to values X/k is called 'changing the scale'. Values of X are measured in units of 1, values X/k in units of k (how many k in X?). The value k in the X system becomes value $k/k = 1$ in the new system. This fact too can sometimes make arithmetic easier. For example, suppose we want the arithmetic mean of the numbers 24, 30, 36, 42 and 48. Dividing each number by 6 gives as new values the numbers 4, 5, 6, 7 and 8. The arithmetic mean of the new values is $(4+5+6+7+8)/5 = 30/5 = 6$, so the mean of the original values is $6 \times 6 = 36$.

When values X are arranged in a frequency distribution the calculation of the arithmetic mean is best done in table form. Consider first a discrete distribution and take as example the distribution of number of bedrooms (Table 2.1, p. 13). Since $\bar{X} = (\Sigma Xf(X))/n$ we need to find $\Sigma Xf(X)$ and n as in Table 3.1. $\Sigma X^2f(X)$ is also given for later use.

Table 3.1

X	$f(X)$	$Xf(X)$	$X^2f(X)$
1	5 670	5 670	5 670
2	15 120	30 240	60 480
3	38 430	115 290	345 870
4	2 520	10 080	40 320
5	1 260	6 300	31 500
	63 000	167 580	483 840

$$\bar{X} = \frac{\Sigma Xf(X)}{\Sigma f(X)}$$

$$= \frac{167\,580}{63\,000}$$

$$= 2{\cdot}66 \text{ bedrooms}$$

We can interpret this by saying that if the total number of bedrooms (167 580) were divided equally among the 63 000 houses each house would have 2·66 bedrooms. This is impossible practically, of course, but theoretically is correct. The arithmetic mean is 2·66, and neither 2 nor 3. When values are discrete the arithmetic mean is often a value that cannot occur.

When finding the arithmetic mean of a grouped distribution we take as values X the mid-points of the classes. Thus open-ended classes have to be closed to enable all mid-points to be found before the arithmetic mean can be calculated. It is with grouped distributions that changing the origin (subtraction) and changing the scale (division) make calculations noticeably easier. Calculations are best done in table form.

As an example consider the distribution in Table 3.2 which relates to travel to work from rural districts to urban districts in 1966 (taken from *Social Trends No. 2*, Government Statistical Office, 1971). Here the variable is the proportion of economically active population in the rural area travelling to an urban area and 413 rural districts appear in the distribution.

In Table 3.2 values X have first been changed to values

Table 3.2

Proportion travelling (as a %)	*Mid-points* X	*No. rural districts* $f(X)$	$Y = X-35$	$Z = \frac{X-35}{10}$	$Zf(X)$	$Z^2f(X)$
Under 10*	5	32	−30	−3	−96	288
10–20	15	61	−20	−2	−122	244
20–30	25	88	−10	−1	−88	88
30–40	35	81	0	0	0	0
40–50	45	71	10	1	71	71
50–60	55	53	20	2	106	212
60 and over*	65	27	30	3	81	243
		413			258	1 146
					−306	
					−48	

* Class 'under 10' closed as '0–10' and class '60 and over' closed as '60–70'.

$Y = (X-35)$, and then these new values have been changed to values $Z = Y/10 = (X-35)/10$. The arithmetic mean $\bar{X}$ is $35+\bar{Y}$, and $\bar{Y} = 10\,\bar{Z}$. Thus to get $\bar{X}$ adjustments are made in reverse order. The last change to values was division by 10, so the first adjustment to the mean of values Z is multiplication by 10. The previous change was subtraction of 35, so the next adjustment to the mean found is addition of 35. (The column $Z^2f(X)$ is also given in Table 3.2 for later use.)

Arithmetic mean of values Z is

$$\bar{Z} = \frac{\Sigma Zf(X)}{n} = -\frac{48}{413} = -0{\cdot}1162 \text{ (unrounded)}$$

Arithmetic mean of values Y is

$$\bar{Y} = 10\,\bar{Z} = -\frac{(10)(48)}{413} = -1{\cdot}162$$

Arithmetic mean of values X is

$$\begin{aligned}\bar{X} &= 35+\bar{Y}\\ &= 35-1{\cdot}162\\ &= 33{\cdot}838\\ &= 33{\cdot}84\% \text{ (correct to 2nd decimal place)}\end{aligned}$$

There is no need to learn a separate formula including changes in origin and scale. The basic formula $\bar{X} = \Sigma Xf(X)/n$ still holds. Changes from X to Y and Z (or any notation the reader likes to use) and the adjustments needed to get back to $\bar{X}$ from $\bar{Z}$ are best remembered as method.

The choice of what number to subtract is arbitrary, but $\Sigma Zf(X)$ is a minimum (in fact zero) when the subtracted number is $\bar{X}$. Thus a useful first step in choosing the number is to guess the value of $\bar{X}$. Next notice that if we choose one of the mid-points as the number to subtract one of the new values is zero, and zero is an easy value to have in calculations. Thus a suitable number to subtract is the mid-point of the class in which we guess $\bar{X}$ lies. This means that our guess at the value of $\bar{X}$ may be fairly rough. Notice that unless the distribution is very skew subtracting a value close to $\bar{X}$ will give a mixture of negative and positive values as new values. This means that $\Sigma Yf(X)$ and $\Sigma Zf(X)$ could either be negative or positive. Notice too that generally we obtain the smaller new numerical values (whether negative, zero or positive) with the larger frequencies, which makes multiplications $Yf(X)$ or $Zf(X)$ easier than multiplications $Xf(X)$.

It is not always possible to adjust values further by division, but when all classes are the same width we can always divide by the common class width and that will result in the most saving. This is one reason why, when distributions have open-ended classes it is sensible to close the classes to make them the same width as the other classes, as far as is possible. In the example above, the two open-ended classes were closed to make all classes of width 10 and we divided values Y by 10. When classes are all the same width we can always change the mid-points into values like -2, -1, 0, 1, 2 and these are the values with which we work when finding the arithmetic mean. We aim to divide by the highest common factor which leaves whole numbers.

The reader will find that with practice he will be able to calculate arithmetic means from grouped distributions without writing down all mid-points, without finding values Y as an intermediate step, and, of course, without finding $\bar{Z}$ and $\bar{Y}$ as intermediate steps. Thus, in the example above, he will omit columns headed 'mid-points X' and headed '$Y = X-35$'. Underneath the table he will write just 'Arithmetic mean of marks X is $\bar{X} = 35+10\,(-48/413)$' and the calculation. He should not, however, omit the headings of the remaining columns.

As a second example of calculation of an arithmetic mean from grouped data, take the distribution of ages of working males in Great Britain (Table 2.2, p. 18). Working is shown in Table 3.3.

Table 3.3

Ages	*Mid-points* X	$f(X)$	$Y = X-35{\cdot}0$	$Z = (X-35)/2{\cdot}5$	$Zf(X)$
15–19	17·5	1 200	−17·5	−7	−8 400
20–24	22·5	1 900	−12·5	−5	−9 500
25–44	35·0	6 340	0	0	0
45–64	55·0	6 020	20·0	8	48 160
65 and over*	67·5	540	32·5	13	7 020
		16 000			55 180
					−17 900
					37 280

* Closed as 65–69.

Referring to Table 3.3 we find that

$$\bar{X} = 35+\frac{(5)\,(37\,280)}{(2)\,(16\,000)}$$

$$= 40{\cdot}825 \text{ years}$$

Although the classes are not all the same width in this distribution we are still able to make a division which saves arithmetic.

Although the arithmetic mean is the hardest of the three averages – mode, median, arithmetic mean – to calculate, and cannot be found diagrammatically, it is the most used of the three. The chief reason for this is that the arithmetic mean has good algebraic properties, and, as is to some extent shown later in this book, it also has what could be termed good statistical properties.

One important property of the arithmetic mean is that if the arithmetic means of several different distributions of similar data are known we can find the arithmetic mean of the combined distribution from the separate means and total frequencies. Suppose the means of the different distributions are $\bar{X}_1$, $\bar{X}_2$, ---, $\bar{X}_k$ and their total frequencies are n_1, n_2, ---, n_k. Now

$$\bar{X}_1 = \frac{\text{total amount in 1st distribution}}{n_1}$$

so total amount in 1st distribution is $n_1\bar{X}_1$. Similarly the total amounts in the other distributions are $n_2\bar{X}_2$, ---, $n_k\bar{X}_k$, so the total amount available when the distributions are combined is $n_1\bar{X}_1 + \text{---} + n_k\bar{X}_k$. The total frequency when the distributions are combined is $n_1 + \text{---} + n_k$. The arithmetic mean of the combined distribution is

$$\frac{\text{total amount in combined distributions}}{\text{total frequency of combined distributions}} = \frac{n_1\bar{X}_1 + \text{---} + n_k\bar{X}_k}{n_1 + \text{---} + n_k}$$

It is a 'weighted average' of the individual means, or a 'mean of means', the weights being the total frequencies of the separate distributions.

This fact saves greatly on arithmetic when we need means of both separate and combined distributions. The mode or median of a combined distribution cannot be found from the modes or medians of the separate distributions.

Unlike the mode and the median, the arithmetic mean is calculated from all the values in the distribution, so that when it seems important that all values are considered the arithmetic mean is the appropriate average. Thus, the arithmetic mean is influenced by atypical extreme values and this can sometimes be a disadvantage. Of the three measures it is the most stable in the sense that changes of grouping in a grouped distribution will not change its value much. Calculating the mean of a grouped distribution is easier and quicker than calculating it from a large number of ungrouped values, and so it is often worth grouping before finding the mean, since the value of a mean found from a grouped distribution is not very different from the mean calculated from the ungrouped data.

When a distribution is symmetrical (exactly, not approximately) and of 'normal' shape, the mode, the median, and the arithmetic mean all have the same value. When a distribution is skewed to the right the mode is less than the median and the median is less than the mean (mean to the right). When a distribution is skewed to the left the mean is less than the median and the median is less than the mode (mean to the left). The greater the skew, the farther apart are the mean, the median, and the mode.

The mean deviation

As a first step in finding a measure of spread to use with the arithmetic mean consider finding the distance of each value X from the mean $\bar{X}$. This distance is $|X-\bar{X}|$ (i.e. modulus of $(X-\bar{X})$. We measure distance as positive even when $(X-\bar{X})$ is negative. As a summary measure we take the arithmetic mean of the distances, that is, $(1/n)\Sigma|X-\bar{X}|f(X)$ where X occurs with frequency $f(X)$. This quantity is called the *mean deviation.*

Note that if we try to take account of the direction of the distance, that is, the sign of $(X-\bar{X})$, the measure would be $(1/n)\Sigma(X-\bar{X})f(X)$ which is always zero, for

$$\frac{1}{n}\Sigma(X-\bar{X})\,f(X) = \frac{\Sigma X f(X)}{\Sigma f(X)} - \bar{X}\frac{\Sigma f(X)}{\Sigma f(X)} \quad \text{writing } n = \Sigma f(X)$$

$$= \bar{X}-\bar{X}$$

$$= 0$$

The method of calculation of the mean deviation for an extremely simple distribution of six values with a mean of 6 is shown in Table 3.4.

Here calculation is easy as the mean is a whole number

Table 3.4

X	$X-\bar{X}$	$\lvert X-\bar{X}\rvert$
1	−5	5
3	−3	3
5	−1	1
6	0	0
10	4	4
11	5	5
36	0	18

$$\bar{X} = \frac{36}{6} = 6 \qquad \text{Mean deviation} = \frac{18}{6} = 3$$

and all the frequencies are 1. With a decimal mean, or with frequencies greater than 1, the calculation becomes tedious. The reader can verify this by changing the value 11 in the above distribution to 12 and calculating the mean deviation of that new distribution, or by finding the mean deviation of the distribution of bedrooms. With grouped frequency distributions the calculation could be even more lengthy. There is no need to write down $(X-\bar{X})$ as well as $|X-\bar{X}|$, though it is useful to check that $\Sigma(X-\bar{X})f(X) = 0$.

Thus, although the mean deviation is easy to understand (as the average distance from the mean), it is not very satisfactory in use and it has no redeeming algebraic or statistical properties. It is therefore not much used.

The variance and standard deviation

The distance of X from $\bar{X}$ is obviously important in measuring spread about $\bar{X}$. We could square all the distances $(X-\bar{X})$ and take the arithmetic mean of the n quantities obtained, that is, find

$$\frac{1}{n}\Sigma(X-\bar{X})^2 f(X)$$

$$= \Sigma(X-\bar{X})^2\frac{f(X)}{n}$$

$$= \Sigma(X-\bar{X})^2 \text{ times relative frequency}$$

where X occurs with frequency $f(X)$. This is called the *variance* of X and is denoted by S^2. The variance is the second moment about the mean (compare moments in mechanics) and is always positive, since it is the mean of positive quantities $(X-\bar{X})^2$.

If X is measured in units, whether of money, weight, time, or something else, the variance will be in 'units squared'. To obtain a measure in units take the square root of the variance, that is find $\sqrt{[(1/n)\Sigma(X-\bar{X})^2 f(X)]}$. This is called the *standard deviation* (abbreviated to S.D.) of X. By convention we take the positive square root.

The formula for the variance can be written in a form more suitable for computation (see Appendix 2). This is variance $= (1/n)\Sigma X^2 f(X) - \bar{X}^2 =$ [average of (X^2)] $-$ [average of $X]^2$ where 'average' means 'arithmetic mean'.

$$\text{S.D.} = \sqrt{\text{variance}}$$

as before. The reader should note carefully what is being squared in each term. Except in a very few cases the computational form of the formula should be used in calculations in preference to the form given defining the variance. Not only is the arithmetic easier using the computational form, but when rounding is involved the answer will nearly always be more accurate, for when it is not practical to find the exact value of $\bar{X}$, every $(X-\bar{X})$ is in error, and squaring $(X-\bar{X})$ magnifies this error. In the computational form of the formula the error in $\bar{X}$ is still there but contributes far less to the final answer.

Consider making every value X smaller by an amount c, that is, changing the origin, as we did when discussing the arithmetic mean. Then the mean of the new values is $(\bar{X}-c)$. The variance of the new values is

$$\frac{1}{n}\Sigma[(X-c)-(\bar{X}-c)]^2 f(X) = \frac{1}{n}\Sigma(X-\bar{X})^2 f(X).$$

But this is the variance of values X, and must be so since the distance of values from the mean is the same whether values are measured from an origin of zero or an origin of c.

Next consider dividing every value X by a fixed constant k, that is, changing the scale. The mean of the new values is $\bar{X}/k$, so the variance of the new values is

$$\frac{1}{n}\Sigma\left(\frac{X}{k}-\frac{\bar{X}}{k}\right)^2 f(X) = \frac{1}{k^2}\times\frac{1}{n}\Sigma(X-\bar{X})^2 f(X) = \frac{1}{k^2}(\text{variance of } X)$$

To get the variance of X we must therefore multiply the variance of the new values by k^2. Looked at in another way, to get the standard deviation of X, multiply the standard deviation of the new values by k. We normally find the variance, or standard deviation, of the new values after changing the origin or scale by using the computational form of the formula.

Finding the variance, or standard deviation, is much like finding the arithmetic mean. For a frequency distribution calculations are best done in table form as in the following examples.

First take the distribution of the number of bedrooms (Table 2.1). To find the variance we need $\Sigma X^2 f(X) = \Sigma X[Xf(X)]$, $\bar{X} = \Sigma Xf(X)$, and n (all given in Table 3.1, p. 49, with the calculation of $\bar{X}$). Notice that $X^2 f(X)$ is always positive.

$$\begin{aligned}
\text{Variance of } X &= \frac{1}{n}\Sigma X^2 f(X) - \bar{X}^2 \\
&= \frac{483\,840}{63\,000} - \left(\frac{167\,580}{63\,000}\right)^2 \\
&= 7{\cdot}68 \quad - \quad 2{\cdot}66^2 \\
&= 0{\cdot}6044 \text{ bedrooms squared} \\
\text{S.D. of } X &= 0{\cdot}78 \text{ bedrooms}
\end{aligned}$$

As a second example take the distribution relating to travel to work from rural districts of Table 3.2 (p. 50). This illustrates changing the origin and scale in finding the variance of a grouped distribution. Here (referring to Table 3.2)

$$
\begin{aligned}
\text{variance of } Z &= \frac{1}{n}\Sigma Z^2 f(X) - \bar{Z}^2 \\
&= \frac{1\,146}{413} - \left(\frac{-48}{413}\right)^2 \\
&= 2{\cdot}7748 - 0{\cdot}0135 \\
&= 2{\cdot}7613 \\
\text{variance of } X &= 100\ (\text{variance of } Z) \\
&= 276{\cdot}13\ \%\ \text{squared} \\
\text{S.D. of } X &= 16{\cdot}62\ \%
\end{aligned}
$$

The variance and standard deviation, particularly the latter, are not easy measures to understand, and they are harder to calculate than the range and the semi-interquartile range, but they are the most used of the measures of spread. One reason for this is that they are appropriate measures of spread to use with the arithmetic mean and the arithmetic mean is the most used measure of position. The other reasons are similar to the reasons why the arithmetic mean is so popular — good algebraic and statistical properties, variances of separate distributions can be combined to give the variance of the combined distribution (see Walpole, 1974, and Wonnacott and Wonnacott, 1972), all values are considered when calculating the variance, and it is fairly stable as a measure.

Coefficient of variation

One further measure of spread which it is appropriate to use with the arithmetic mean is the *coefficient of variation*, abbreviated to C of V.

$$C \text{ of } V = \frac{\text{S.D.}}{\text{arithmetic mean}}$$

and so incorporates two measures which have many points in their favour.

The chief use of the coefficient of variation is in comparing different distributions since it is a dimensionless measure and to some extent removes differences due to differences in the values of the means. Thus, to compare the variability of, say, two income distributions, one in dollars and one in pounds sterling, we would say that the one with the larger coefficient of variation is the more variable. On the other hand two distributions which obviously have the same spread (by examination of frequencies) may have different coefficients of variation just because their means are different and this is a distinct disadvantage.

Exercises

1 Refer to chapter 2, questions 5 and 6, and for each set of data, calculate the mode, the median, the arithmetic mean, and the range.

2 The following table gives the frequency distribution of the lengths of surnames of account holders at a large shop. Find the mean length of surname, the median length, and the modal length.

Do you think these three numbers alone would enable the accounts department to decide how many spaces to allow for the surname on account cards? If not, why not?

No. letters in surname	*Frequency*
3	8
4	48
5	104
6	130
7	106
8	64
9	28
10	7
11	5
	500

3 Refer to the grouped distribution obtained from chapter 2, question 7, and

- i estimate the mode diagrammatically
- ii calculate the median from the grouped distribution and give also the exact value of the median
- iii estimate the number of people who obtained at least 60 but less than 75 correct answers
- iv find the arithmetic mean
- v explain how to find the mean deviation.

4 In a survey of 500 students who had been at College for at least four years a question was asked about readership of the college newspaper. The newspaper is issued in each of the twelve months of the year and students were shown every copy issued in the last four years and asked if they had seen it previously. The frequency distribution below shows the number of students who had seen different numbers of issues.

No. of issues seen	*No. of students*
0–6	140
7–13	75
14–20	25
21–27	20
28–34	35
35–41	90
42–48	115
	500

Find the mode or modes, the median, and the arithmetic mean. Which measure do you think best describes this distribution, and why? Find the standard deviation and the semi-interquartile range.

5 Refer to chapter 2, question 8, and find the median and semi-interquartile range of the distribution
(*a*) diagrammatically, (*b*) by calculation.

6 The distribution below shows the times in minutes taken to complete a car treasure hunt by 300 persons.

Time (in minutes)	*No. of persons*
40–45	35
45–50	36
50–55	39
55–60	42
60–65	60
65–70	50
70–75	38
	300

i Find the mode

ii calculate the median and semi-interquartile range

iii estimate how many persons took (*a*) between 41 and 53 minutes, (*b*) more than 67 minutes.

7 The following distribution is of examination marks obtained in two different subjects. Both examinations were taken at approximately the same time. Compare the distributions by calculating for each the arithmetic mean, the standard deviation, and the coefficient of variation.

	Frequency of mark	
Range of marks	*Subject 1*	*Subject 2*
30–35	1	1
35–40	4	2
40–45	10	3
45–50	12	12
50–55	50	32
55–60	33	45
60–65	16	20
65–70	3	8
70–75	1	7
	130	130

8 Distribution A below is the age distribution of persons living in a new town. Distribution B is the age distribution of persons living in an old urban area. What shape is each distribution? Compare the distributions by calculating the arithmetic mean and variance of each.

Distribution A		*Distribution B*	
Age (in years)	*Frequency*	*Age (in years)*	*Frequency*
0–10	190	under 10	10
10–20	240	10–20	10
20–30	200	20–30	60
30–40	170	30–40	120
40–50	130	40–50	170
50–60	50	50–60	200
60–70	10	60–70	250
70 and over	10	70–80	180
	1 000		1 000

4
Probability

The concept and subject of probability is very important in statistics, and, as stated in chapter 1, where probability theory was described as being theory relating to the chance or likelihood of events, statistics began with probability at a very early date. In this chapter we discuss how we can measure chance and then develop some of the elementary theory of probability. We shall use the consequences of the theory in later chapters, particularly in sections relating to distributions.

In statistics we deal with events that are subject to chance variations. We might say that events follow laws involving elements of chance, that is, probabilistic laws, instead of deterministic laws where the outcome of an event is certain. Thus, any prediction of the outcome of an event or experiment in statistics needs to be qualified with the chance that it will occur.

To illustrate how outcomes are subject to chance variations even when the result seems certain, consider the problem of dividing the circumference of a circle by the diameter of the circle. If the radius of the circle is of length r, then the diameter is of length $2r$, and the circumference is of length $2\pi r$ where π is a mathematical constant. Thus

$$\frac{\text{length of circumference}}{\text{length of diameter}} = \frac{2\pi r}{2r} = \pi$$

Therefore, if we measure the circumference and diameter of a circle such as the cross-section of a cylindrical tin, and divide the length of the circumference by the length of the diameter, we are certain to get an estimate of the number π.

Now consider some practical details. We might choose a tin whose cross-section is not perfectly circular, although it appears to be so. In measuring the circumference of the tin by say, seeing what length of cotton is needed to go round the tin, the cotton might slip slightly so that we measure a wrong length. In measuring the diameter we might by mistake measure a line which does not pass through the centre of the circle, although it almost does so. Our measurement, in itself, could well be inaccurate, either because our measuring apparatus is not sufficiently finely calibrated, or because our reading of it is slightly in error – or for both these reasons. Our division of the one length by the other might be in error. Thus chance is affecting the result in various ways – there is a chance element involved in choosing the tin, chance plays a part in deciding whether we measure the circumference and diameter or some other two lengths, chance affects the accuracy of the measurements and the arithmetic. The reader might be able to add to this list. The outcome of the experiment is thus subject to chance variations, some of which are measurable to some extent.

Consider repeating the experiment several times, possibly with different tins, with different experimenters, and even different measuring apparatus. For each experiment the result of dividing the length of the circumference by the length of the diameter should be the number π. However, the outcomes of the experiment are subject to chance, and very few results will be equal to π, although most will be close to π. We describe the outcomes of the experiment as being a statistical regularity. Statistics enables us to analyse such results and sort out the different variations. To do this we need to consider the chance element more carefully, and probability theory enables us to do this.

It is also possible to have an experiment where the outcome is in no way certain, although the set of outcomes of the experiment is of statistical regularity. For example, one-fifth of the tins used in the previous experiment might have the same coding marked on their lids. If a new experiment consists in choosing a tin and recording the coding we cannot be certain what the coding will be for any one tin. However, if a set of results is examined, a pattern is seen, namely that about a fifth of the tins have the same coding. A similar example in natural science concerns the movement of particles in a substance. The collisions of particles do not follow a deterministic law, but records of collisions show patterns of statistical regularity, for example, in the number occurring per unit of volume in a unit of time. In the social sciences there are very few deterministic laws in the sense that in social science it is not usually possible to say what the outcome of a particular experiment or situation will be, but in long-term or large-scale studies of the situation definite patterns emerge and these suggest laws. Studies of incomes, strikes, social mobility, suicides, and traffic accidents are just a few examples from social science where probabilistic laws can be seen to be operating. Such situations can be investigated mathematically (see Bartholemew, 1967).

We are all accustomed to effects of chance on our lives and often refer to having good or bad luck, or to fate – the fortune teller's way of dealing with probability. As familiar examples of probability consider the following. In raffles every number is as likely to be the winning number as any other. Premium Bond winners are determined on a similar principle – ERNIE is simply a kind of raffle. In betting, as for example on the outcome of a horse race, we talk about the 'odds' for a particular horse winning. Odds, as we shall see, are derived from probability. Allocation of prizes in casinos is determined by consideration of probability. Measurement of the risk involved in taking a drug, or in smoking, or in travelling by different modes of transport is by probability. Closely connected with risks are insurance premiums. If the risk of a disaster is high the premium to insure against it is high. Insurance premiums are calculated using probability theory.

The measurement of probability

We need to define probability and decide how to measure it and discuss four different approaches here.

(1) *Equally likely approach*

The *equally likely* approach to probability is generally the easiest to apply. Suppose we have an experiment which has n equally likely outcomes, and suppose that r of the n outcomes give the result that event E happens, that is r of the n outcomes are favourable to event E, or we could say there are r out of n chances of getting E. Then we say that the probability of E, often written probability (E) or prob (E), is r/n. For example, suppose that the experiment consists in tossing an ordinary six-faced die. If the die is accurately made and we assume that no such event as losing the die occurs, then the experiment has six equally likely outcomes – any of the numbers 1,2,3,4,5,6 falls uppermost. Three of these outcomes are favourable to the event 'odd number uppermost' so the probability of an odd number is said to be 3/6 or 1/2. Note that this approach suffers from the disadvantage of defining probability in terms of equally likely, that is, equally probable, events. In other words it defines probability in terms of probability.

The reader should be careful not to apply equally likely theory indiscriminately. If an experiment can be described as having *n* outcomes, these outcomes are not necessarily equally likely. For example, we could say that tossing a die gives two results – an odd number or an even number. These events are equally likely as we know from experience, or using the equally likely definition of probability and considering the six outcomes as above we find

prob (odd number) = 3/6 = 1/2 as above
and prob (even number) = 3/6 = 1/2

Thus we can say the experiment has two equally likely outcomes – an odd number or an even number. One of the outcomes favours the event 'odd number uppermost' so the probability of an odd number is 1/2.

However, if we say that tossing the die gives two results – a number divisible by three or a number not divisible by three – then these two results are not equally likely. For, considering again the experiment as having six equally likely outcomes, two of the outcomes favour the result 'number divisible by three' whereas four of the outcomes favour the result 'number not divisible by three'. Thus

prob (number divisible by three) = 2/6 = 1/3
whereas prob (number not divisible by three) = 4/6 = 2/3

In this case we have described the experiment as having two outcomes which are not equally likely.

(2) *Relative frequency approach*

The *relative frequency* approach to probability is a natural approach. Suppose we have an experiment which has *k* outcomes, which might or might not be equally likely, and suppose we repeat the experiment N times and record the number of times each outcome occurs, that is, the frequency of each outcome. If outcome E occurs with frequency f, then the relative frequency of E is f/N. If we continue repeating the experiment and record the relative frequency of E at various stages, then we expect the situation to stabilize and the value of the relative frequency to settle down so that it will not be changed much by additional repetitions of the experiment. We define the probability of E to be this limiting value of the relative frequency of E as the number of times the experiment is performed increases indefinitely. Mathematically we write

$$\text{prob}\,(E) = \lim_{N \to \infty} \frac{f}{N}$$

which is read as

prob (E) = limit as N tends to infinity of f/N.

As illustration of this approach consider again the example of the six-faced fair (that is, accurately made) die. If we tossed the die just six times we should not be unduly surprised if none of the six results was 'side two uppermost'. Nor should we be unduly surprised if, say, two of the six results were 'side two uppermost'. In the first case the relative frequency of side two is $0/6 = 0$. In the second case the relative frequency of side two is $2/6 = 1/3$. The equally likely approach tells us that the probability of side two is $1/6$, and of course we might find that in six repetitions of the experiment the relative frequency of side two was $1/6$, but we are prepared for other relative frequencies. Reasoning tells us that if we toss the die N times we expect $1/6$ of the N results to be 'side two uppermost', so if we toss it sixty times, say, we expect ten of the results to be 'side two', giving the relative frequency of side two as $10/60 = 1/6$. However, we are fairly well prepared for the number of

F

times we get side two to be some other number fairly close to 10 such as 8,9,11, or 12 say.

Table 4.1 gives a set of fictitious results which we could well get if we experimented tossing a fair six-faced die different numbers of times and recorded the frequency of side two. In the limit, if we could toss the die an infinite number of times and find the frequency of side two we would find that the relative frequency of side two was exactly 1/6. Now 1/6 = 0·167 rounded to the third decimal place, and the relative frequencies shown in Table 4.1 are getting closer to 0·167 as the number of tosses of the die increases.

Table 4.1

No. of tosses	*Frequency of side two*	*Relative frequency of side two (rounded to 3rd decimal place)*
6	0	0/6 = 0·000
12	1	1/12 = 0·083
36	4	4/36 = 0·111
60	12	12/60 = 0·200
120	18	18/120 = 0·150
600	102	102/600 = 0·170
6000	998	998/6000 = 0·166

It is a useful exercise to check on the relative frequency approach by tossing a die or a coin a large number of times and recording relative frequencies at various stages, similarly to Table 4.1, to see just how rapidly the relative frequency does approach the value we expect. Note, however, that since we do not know at the outset if the die or coin is fair, the relative frequency could well approach a value slightly different from that expected. In addition there is the practical difficulty that continued handling of the die or coin will wear it and possibly change the different probabilities associated with it.

In this example we have subconsciously appealed to equally likely theory and have decided that the relative frequency approach would give the same values of probability. Needless to say the equally likely approach is quicker and neater in use.

(3) *Axiomatic approach*

The most satisfactory approach to probability from the theoretical point of view is the *axiomatic* or *mathematical approach*. This defines a measure called probability and lays down a set of axioms which this measure must follow. The theory depends greatly on measure theory and in consequence on the mathematical theory of sets, and as such is a branch of pure mathematics. This approach does not give numerical values of probabilities, but the probabilities given by the equally likely and relative frequency approaches satisfy the axioms of the axiomatic approach, and it does enable rigorous proof of what might be termed the laws of probability. It has also facilitated the development, not only of many statistical methods with practical applications, but also of applications within the field of probability itself. To mention a few of these – probabilistic approaches to social mobility mentioned earlier; studies of queues in general, and in special situations such as aircraft waiting to land or leave, machines lying idle while waiting to be serviced, and demand on telephone lines; studies of the spread of epidemics or diseases; and studies relating to radioactive particles in high energy physics – are all examples of the development of axiomatic probability theory with reference to particular practical problems.

(4) *Subjective or personal approach*

The *subjective* or *personal approach* to probability can be thought of loosely as one's own guess at the probability of an event. It is used mainly when the equally likely and relative frequency approaches break down, so typically relates to 'once only' type of events, for example, the probability that a certain building will be burnt down tonight, or the probability that a particular person will lose a limb in a car accident tomorrow. Obviously even guesses at probabilities such as these tend to be based on experience, which generally means the relative frequency of similar events. One way of assessing subjective probabilities is by finding out what risks one would take on the event occurring or not occurring, say in monetary terms. It is possible to develop theory relating to subjective probability, but this approach should only be used in special situations.

Numerical measurement

When we need the numerical value of a probability we get this by using the equally likely approach where possible. In other cases where we have data in the form of frequency distributions, we estimate the probability of X (or for grouped data the probability of the class whose mid-point is X) by its relative frequency $f(X)/\Sigma f(X)$.

In a sense the measure of probability is an ideal. We can postulate its existence and require it to satisfy certain axioms which are in agreement with our ideas about what a measure of probability should be like, but the only value we can find for it is an experimental value which has to stand in for the true unknown value. We can quote probabilities as fractions, as percentages, or as decimals, whichever is most convenient.

Events and the sample space

In probability theory we refer to events. If 'E' is an event we say that 'not E' is an event also, for example, in tossing a die getting the result 'side six uppermost' is an event and so also is getting the result 'any side except side six uppermost'.

Events are said to be *mutually exclusive* if they cannot happen together, for example, in a single toss of the die we cannot get the event 'side three uppermost' and the event 'side four uppermost' at the same time, so these events are mutually exclusive. On the other hand we can get the events 'even number' and 'side four uppermost' at the same time, so these events are not mutually exclusive. Events are said to be *exhaustive* if at least one of them must occur, for example, events 'E' and 'not E' are exhaustive, in tossing a die the six events, sides 1,2,3,4,5,6 uppermost are exhaustive.

In any situation there are a number of events which might happen, and a *sample space* consists of a set of mutually exclusive and exhaustive events. For example, when tossing a die the events which can happen, assuming the die is not lost, are sides 1,2,3,4,5,6 uppermost, and this set of events is a sample space. There is often more than one sample space in terms of mutually exclusive and exhaustive events. For example, two other sample spaces for tossing a die are the space consisting of events 'even number' and 'odd number', and the space consisting of 'side six uppermost' and 'any side except side six uppermost'. The reader will be able to think of other sample spaces for this experiment. The sample space to choose is the one most appropriate for the problem, or if this is not obvious or seems unhelpful, the sample space giving the most scope. In many cases there is no need to make conscious reference to a sample space at all.

A *simple event* is one which belongs to the sample space. A *compound event* is one which is in some way a combination of events belonging to the sample space, for example, the events 'getting an even number' and 'getting either side one or side five' are compound events for the first space given above for the die.

We can represent events and sets of events by means of Venn diagrams as shown in Figures 4.1. The rectangle represents the sample space.

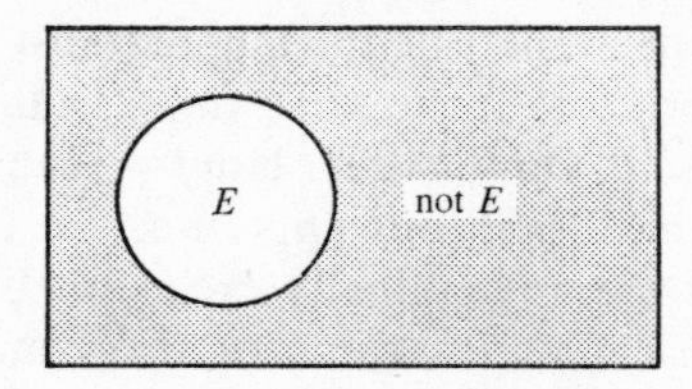

Figure 4.1 (i)

The circle represents event 'E', the shaded area represents event 'not E'.

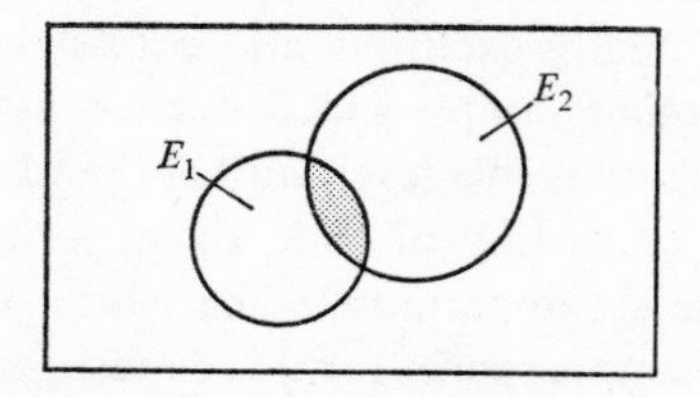

Figure 4.1 (ii)

One circle represents event E_1 the other represents event E_2. The shaded area represents the event 'both E_1 and E_2 together'. When E_1 and E_2 are mutually exclusive there is no area of overlap.

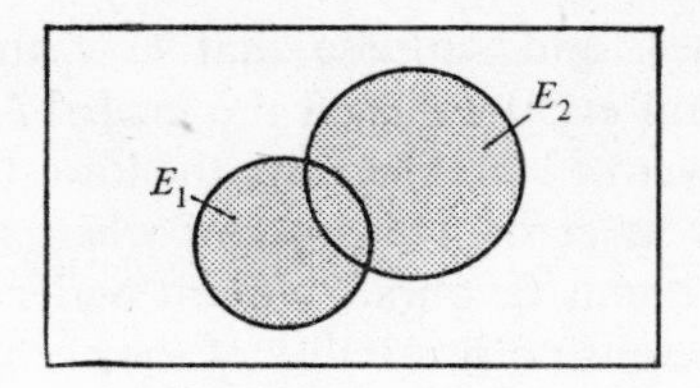

Figure 4.1 (iii)

One circle represents E_1 the other represents E_2. The shaded area shows the event 'either E_1 or E_2' which can alternatively be described as the event 'at least one of E_1 or E_2'.

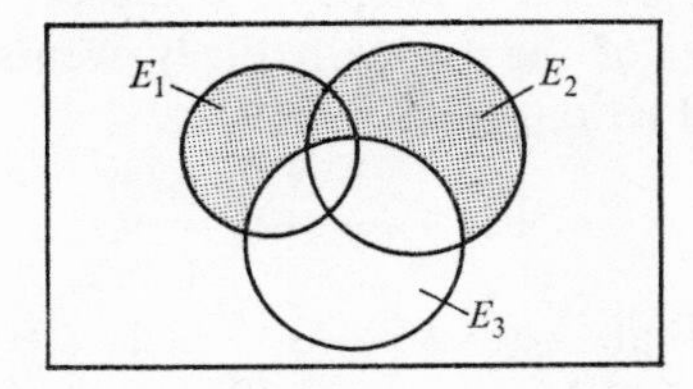

Figure 4.1 (iv)

One circle represents E_1, one E_2, and one E_3. The shaded area shows the event 'either E_1, or E_2, but not E_3'.

Properties and simple rules of probability

In discussing properties and rules of combination of probability we use the equally likely approach since this is mathematically the least demanding. The rules are the same as those obtained by the axiomatic approach. We shall adapt Venn diagrams to illustrate the argument leading to the rules.

Consider, as before, an experiment which has n equally

likely outcomes, and suppose that r of the n outcomes favour a general event E. Then the probability of E is r/n. Since r can never be less than zero, the lowest possible value of probability is zero. This occurs when none of the n equally likely events favours the event under consideration. In this case there is no possibility of the event occurring, so we say prob (impossible event) = 0. Since r can never be larger than n, the highest possible value of probability is one. This occurs when each of the n equally likely events favours the event under consideration. In this case the event must occur so we say prob (certain event) = 1. Probability may take any value between 0 and 1.

The sum of probabilities of a set of mutually exclusive and exhaustive events is one. For if the events are E_1, E_2, ... , E_k and r_1 of the n equally likely events favour E_1, r_2 favour E_2, and so on, then

$$r_1+r_2+\text{-----}+r_k = n$$

$$\text{so that } \frac{r_1}{n}+\frac{r_2}{n}+\text{-----}+\frac{r_k}{n} = 1$$

that is prob (E_1)+prob (E_2)+-----+prob (E_k) = 1

In particular prob (E)+prob (not E) = 1

This relation is of use when we need prob (E) but it is easier to find prob (not E).

The either/or or additive law

Consider a sub-set of the mutually exclusive events E_1, E_2,---, E_k considered above, say E_1, E_2,---, E_j. Then $(r_1+r_2+\text{----}+r_j)$ events favour the event 'either E_1 or E_2 or --- or E_j' which we can also describe as the event 'at least one of E_1, E_2,---, E_j'. Thus

prob (either E_1 or E_2 or. . . . . or E_j)

$$= (r_1 + r_2 + \cdots + r_j) \,/\, n$$

$$= \frac{r_1}{n} + \frac{r_2}{n} + \cdots + \frac{r_j}{n}$$

$$= \text{prob}\,(E_1) + \text{prob}\,(E_2) + \cdots + \text{prob}\,(E_j)$$

As an example, consider an ordinary pack of 52 playing cards and suppose that the experiment consists in drawing one card from the pack at random, that is drawing the card in such a way that each of the 52 cards is as likely to be drawn as any other. Since there are 4 kings in the pack, prob (king) = 4/52. Similarly since there are 4 queens in the pack, prob (queen) = 4/52. The events 'king' and 'queen' are mutually exclusive. Therefore

$$\begin{aligned} &\text{prob (either king or queen)} \\ &= \text{prob (king)} + \text{prob (queen)} \\ &= 4/52 + 4/52 \\ &= 2/13 \end{aligned}$$

When events are not mutually exclusive more care is needed since straight summing includes double counting. Consider the case of two events E_1 and E_2 which can occur together. Suppose that r_1 of the n equally likely events favour E_1, r_2 favour E_2, and r_3 favour the occurrence of both E_1 and E_2 as shown in Figures 4.2. In the diagrams one circle represents E_1 and contains r_1 events favourable to it, and the other circle represents E_2 and contains r_2 events favourable to it. The area of overlap of the two circles represents the event 'both E_1 and E_2' and contains the r_3 events favourable to it. In the second diagram sections are labelled with the number of events contained in them. This is a more useful and clearer presentation.

The number of events favourable to the event 'either E_1 or E_2' is $r_1 + r_2 - r_3$ which is the total number of events shown in Figure 4.2. Notice that the sum $r_1 + r_2$ includes the r_3

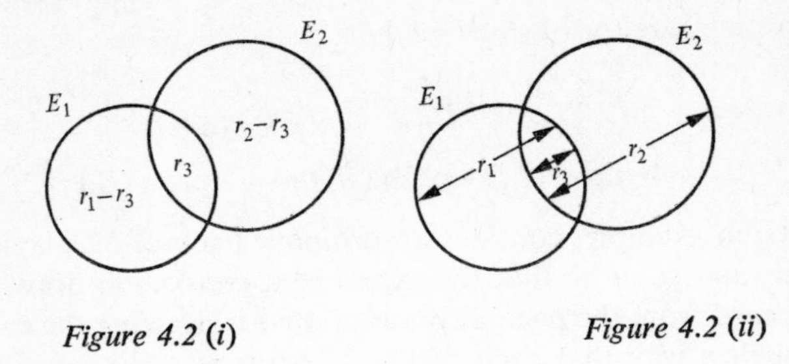

Figure 4.2 (i) *Figure 4.2 (ii)*

events favourable to both E_1 and E_2 twice. Thus in this situation we find

$$\text{prob (either } E_1 \text{ or } E_2)$$

$$= (r_1+r_2-r_3)/n$$

$$= \frac{r_1}{n}+\frac{r_2}{n}-\frac{r_3}{n}$$

$$= \text{prob } (E_1)+\text{prob } (E_2)-\text{prob (both } E_1 \text{ and } E_2)$$

For example, in the pack of cards experiment

prob (either king or a heart)

= prob (king)+prob (heart)−prob (king of hearts)

= 4/52+13/52−1/52

= 4/13

(Note that the formula above covers the case of mutually exclusive events E_1 and E_2 for then $r_3 = 0$ and prob (both E_1 and E_2) is zero.)

The extension of the rule to a set of events which are not mutually exclusive involves fairly complicated mathematical symbolism. In the case of three events the reader will find that reference to a Venn diagram will enable him to find the necessary probability. Similarly such probabilities as prob (exactly one of E_1 and E_2) can be found from Venn diagrams. There are exercises on this at the end of the chapter.

Conditional probability

Sometimes we can use information about an event in order to find a particular probability in which we are interested, thereby improving our estimate of the value by using this additional information. For example, suppose in drawing a card from the pack of 52 we want the probability that the card is the ace of spades. Then, with no information, the probability that the card is the ace of spades is 1/52. However, if we know that the card is an ace, the probability that it is the ace of spades is 1/4 since there are 4 aces, all equally likely, and 1 of the 4 favours the event 'ace of spades'. The probability of 1/4 is said to be the probability that the card is a spade, that is, it is the ace of spades, *conditional* on the event, or given, that it is an ace. In finding this conditional probability we have referred to a reduced sample space – the events ace of spades, ace of diamonds, ace of hearts, and ace of clubs.

In other situations we can use information about previous events in order to find a particular probability in which we are interested. For example, suppose that we have drawn one card from the pack of 52 and then we draw a second card without replacing the first card so that our second draw consists in taking one card from a pack of 51 cards. If we are interested in the probability that the second card we draw is the ace of spades, knowing whether or not the first card was the ace of spades will clearly be helpful. Here the probability that the second card is the ace of spades is *conditional* on the result of the first draw.

In this example with a pack of cards it is easy to find

conditional probabilities since we can count up the number of equally likely possibilities at any stage. In a more general situation we might not be able to do this. Let us consider the relation between conditional probabilities and other probabilities. Suppose we have the events E_1 and E_2 discussed in the previous section on the additive law. The situation was illustrated in Figure 4.2. If we know that E_1 has occurred, the probability of E_2 conditional on E_1, which we can describe as the probability of E_2 given E_1, and write as prob $(E_2|E_1)$, is r_3/r_1 since r_3 of the r_1 equally likely events favouring E_1 also favour E_2. Now

$$\frac{r_3}{r_1} = \frac{r_3}{n} \times \frac{n}{r_1}$$

But r_3/n is the probability of 'both E_1 and E_2' and r_1/n is the probability of E_1, so we have the relation

$$\text{prob}\ (E_2|E_1) = \frac{\text{prob (both } E_1 \text{ and } E_2)}{\text{prob}\ (E_1)}$$

In the cards example E_1 is an ace, E_2 is a spade. The probability that the card drawn is both an ace and a spade, that is, is the ace of spades, is 1/52. The probability that it is an ace is 4/52. Substituting these values in the formula above gives

$$\text{prob (spade|ace)} = \frac{1/52}{4/52} = \frac{1}{4}$$

as before.

As a second example suppose that E_1 is the event 'spin dryer mechanism of a washing-machine defective' and E_2 is the event 'washing-machine passed as perfect by quality control section of the factory'. Suppose that 1 per cent of the machines manufactured have defective spin dryer mechanisms, that is prob $(E_1) = 1/100$. Suppose that the probability of a machine having a defective spin dryer and being passed as perfect by quality control is 1/5 000, that is, prob (both E_1 and E_2) = 1/5 000.

Then

$$\text{prob (washing machine passed as perfect/spin dryer defective)}$$

$$= \frac{\text{prob (machine passed as perfect and spin dryer defective)}}{\text{prob (spin dryer defective)}}$$

$$= \frac{1/5\,000}{1/100}$$

$$= 1/50$$

Notice that

$$\text{prob}\ (E_1|E_2) = \frac{\text{prob (both } E_1 \text{ and } E_2)}{\text{prob}\ (E_2)}$$

$$\text{so prob (both } E_1 \text{ and } E_2) = \text{prob}\ (E_1)\ \text{prob}\ (E_2|E_1)$$
$$= \text{prob}\ (E_2)\ \text{prob}\ (E_1|E_2)$$

The rule for conditional probabilities can be generalized to more than two events. It works out as a repeated application of the rule for two events.

The multiplication law and independent events

If knowledge of the occurrence of E_1 does not add to our knowledge about E_2 so that the occurrence of E_2 does not depend in any way on the occurrence of E_1 then

$$\text{prob}\ (E_2|E_1) = \text{prob}\ (E_2)$$

and

$$\text{prob (both } E_1 \text{ and } E_2) = \text{prob}\ (E_1)\ \text{prob}\ (E_2)$$

The events E_1 and E_2 are said to be statistically *independent* of one another.

As an example consider tossing a coin with faces labelled head (H) and tail (T). If the coin is tossed twice there are four possible and equally likely outcomes:

	1st toss	*2nd toss*
i	head	head
ii	tail	tail
iii	head	tail
iv	tail	head

Thus the probability of both a head on the first toss and a head on the second toss is 1/4. In a single toss of the coin there are two equally likely outcomes – head or tail – so the probability of a head on a single toss is 1/2, whatever the number of the toss. Thus

$$\text{prob}\ (H \text{ on 1st and } H \text{ on 2nd}) = 1/4$$

and

$$\text{prob}\ (H \text{ on 1st}) \times \text{prob}\ (H \text{ on 2nd}) = 1/2 \times 1/2 = 1/4$$

so the events 'head on 1st toss' and 'head on 2nd toss' are statistically independent. (Notice that if we describe the sample space as three events – (i) two heads, (ii) two tails, (iii) one head and one tail – these three events are not equally likely.)

In many situations we know, or are able to assume, the independence of events. The multiplication rule can be extended as follows. If events $E_1, E_2, \ldots, E_j$ are independent then

$$\begin{aligned} &\text{prob}\ (E_1 \text{ and } E_2 \text{ and } \ldots \text{ and } E_j) \\ &= \text{prob}\ (E_1)\ \text{prob}\ (E_2) \ldots \text{prob}\ (E_j) \end{aligned}$$

Bayes's Theorem

The Reverend Thomas Bayes (1702–61) was an English philosopher who had an interest in probability theory. The theorem which bears his name is concerned with twisting conditional probabilities the other way round, that is, given probabilities of the form prob $(F|E_i)$ to find conditional probabilities of the form prob $(E_i|F)$. We illustrate this by means of an example.

Suppose that 35 per cent of the cups in a canteen are white, 25 per cent are green, and 40 per cent are blue. Suppose further that 10 per cent of the white cups are cracked as are 5 per cent of the green and 7 per cent of the blue. A man buys a cup of tea and finds that the cup is cracked. What is the probability that the cup is white?

Call the event 'white cup' E_1, the event 'green cup' E_2, the event 'blue cup' E_3, and the event 'cracked cup' F. We know

$$\begin{aligned}
\text{prob}\ (E_1) &= 0{\cdot}35 \\
\text{prob}\ (E_2) &= 0{\cdot}25 \\
\text{prob}\ (E_3) &= 0{\cdot}40 \\
\text{prob}\ (F|E_1) &= 0{\cdot}10 \\
\text{prob}\ (F|E_2) &= 0{\cdot}05 \\
\text{prob}\ (F|E_3) &= 0{\cdot}07
\end{aligned}$$

and we can immediately find

$$\begin{aligned}
\text{prob}\ (E_1 \text{ and } F) &= \text{prob}\ (E_1)\ \text{prob}\ (F|E_1) \\
&= 0{\cdot}35 \times 0{\cdot}10 = 0{\cdot}0350 \\
\text{prob}\ (E_2 \text{ and } F) &= \text{prob}\ (E_2)\ \text{prob}\ (F|E_2) \\
&= 0{\cdot}25 \times 0{\cdot}05 = 0{\cdot}0125 \\
\text{prob}\ (E_3 \text{ and } F) &= \text{prob}\ (E_3)\ \text{prob}\ (F|E_3) \\
&= 0{\cdot}40 \times 0{\cdot}07 = 0{\cdot}0280
\end{aligned}$$

We want prob $(E_1|F)$ and know that

$$\text{prob}\ (E_1|F) = \frac{\text{prob}\ (E_1 \text{ and } F)}{\text{prob}\ (F)}$$

by rules relating to conditional probability. Consider the event F. Either we get F occurring with E_1, that is the cup is both cracked and white, or it is cracked and green, or it is cracked and blue, and these events are mutually exclusive. Therefore:

$$\begin{aligned}\text{prob}(F) &= \text{prob (either '}F \text{ and } E_1\text{' or '}F \text{ and } E_2\text{' or '}F \text{ and } E_3\text{')}\\ &= \text{prob}(F \text{ and } E_1) + \text{prob}(F \text{ and } E_2) + \text{prob}(F \text{ and } E_3)\\ &= 0{\cdot}0350 + 0{\cdot}0125 + 0{\cdot}0280\\ &= 0{\cdot}0755\end{aligned}$$

$$\text{Thus prob}(E_1|F) = \frac{0{\cdot}0350}{0{\cdot}0755} = \frac{70}{151}$$

This problem is illustrated diagrammatically in Figure 4.3. The hatched area represents the cracked cups and this is the sample space of interest given the information that the cup is cracked. The probability we require is the ratio of the area representing cracked white cups to the area representing cracked cups. This is exactly the ratio found previously, but the reader may prefer to argue from the diagram rather than rules of probability.

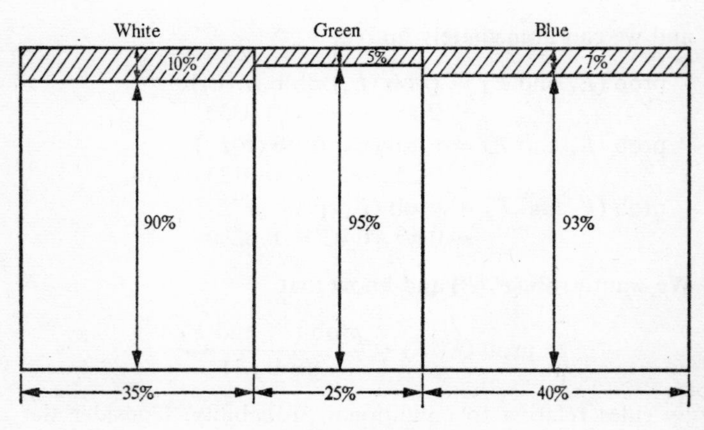

Figure 4.3

In general if there are j mutually exclusive events E_i instead of three as here, Bayes's theorem says

$$\text{prob}\,(E_k/F) = \frac{\text{prob}\,(E_k)\,\text{prob}\,(F|E_k)}{\sum_{i=1}^{j} \text{prob}\,(E_i)\,\text{prob}\,(F|E_i)}$$

(In the worked example $k = 1$ and $j = 3$.) It is better to substitute values in this formula direct in case there are any cancellations, rather than work out probabilities of the type prob (E_i and F) as an intermediate step, as was done in the example.

We can describe the prob (E_i) as prior probabilities – probabilities known before anything happens, and the prob ($E_i|F$) as posterior probabilities – probabilities found after something has happened and using experimental evidence. The so-called 'Bayesian statistics' places great stress on using additional evidence in analysis. This is often in the form of prior probabilities (which might be subjective).

Odds

Sometimes probabilities are measured as odds. If prob (E) is p, then prob (not E) is $(1-p)$. The odds for E are $p/(1-p)$ and the odds against E are $(1-p)/p$. For example, suppose that a particular horse is thought to have a 1 in 50 chance of winning a race. Then prob (wins) = 1/50, prob (loses) = 49/50. The odds for him winning are (1/50)/(49/50) = 1 to 49. The odds against him winning are 49 to 1. The layman might well measure subjective probabilities in terms of odds.

Exercises

1 An ordinary 6-faced fair die is tossed twice. Assuming that the two tosses are independent, find the probability that

i the first toss gives an even number, the second an odd

ii the first toss gives a number divisible by 3, the second is side 4
iii the first toss gives a number not divisible by 3, the second is side 4.

2 One card is drawn at random from an ordinary pack of 52 playing cards. Find the probability that

i it is spades
ii it is an ace
iii it is the ace of spades
iv it is an ace, given that it is spades
v either it is an ace or it is spades
vi either it is the ace of spades or the ace of clubs
vii either it is the ace of spades or the king of spades
viii given that it is an ace, either it is spades or it is clubs
ix given that it is spades, either it is the ace or it is the king.

3 A fair coin is tossed three times. Assume that the tosses are independent of one another. Identify the faces of the coin as heads and tails and find the probability that

i no toss results in head
ii there is at least one head in the three tosses
iii there is exactly one head in the three tosses
iv either there are exactly two heads in the three tosses or exactly two tails
v the first throw is a head
vi the first throw is a head and the last throw is a tail
vii either the first throw is a head or the last throw is a tail
viii the last throw is a tail given that the first throw is a head.

4 A box contains 3 black balls and 5 green balls. A ball is selected at random. If the ball drawn is green it is replaced and two additional green balls are put in the box. If the ball drawn is black, it is not replaced in the box and no additional balls are added. A second ball is then drawn from the box. What is the probability that it is green?

5 Four cubes labelled A,B,C,D are placed in a box. Experiment 1 consists in taking out one cube at random, replacing it, and then taking out a second cube at random. Experiment 2 consists in taking out one cube at random, and then taking out a second cube at random without replacing the first cube. For each experiment find the probability that the second cube is cube B:
(*a*) when the first cube taken out is cube A, (*b*) when the first cube taken out is cube B, (*c*) when the label on the first cube removed is not known.

6 On a housing estate of 1 000 households it is found that 800 households have a refrigerator, 600 households have a washing-machine, and 102 have a deep freeze. Sixty per cent of the households with refrigerators also have washing-machines, but only 9 per cent of households with refrigerators also have deep freezes. Forty-five households have all three appliances: refrigerator, washing-machine, and deep freeze. Twenty households have a washing-machine and a deep freeze but no refrigerator.

If a household is chosen at random, what is the probability that of the three appliances, refrigerator, washing-machine, deep freeze, it has

i none
ii exactly one
iii exactly two?

7 A sample survey is to be done in a town where 40 per cent of the houses are terrace houses, 30 per cent were built before 1900 and 10 per cent have no running water. Half of the terrace houses were built before 1900 and a quarter of these have no running water. In all, a third of the houses built before 1900 have no running water.

What is the probability of obtaining in the sample a terrace house which was built after 1900 and has no running water? What is the probability of obtaining a house built before 1900 which is not terraced and has running water? What is the probability of obtaining a terrace house which has running water? What is the probability of obtaining either a terrace house or one built before 1900?

8 Refer to chapter 2, question 5 (and chapter 3, question 1) and find the probability that (*a*) if a child is chosen at random it comes from a family with 4 children, (*b*) if a family is chosen at random it has 4 children in it.

9 There is a 1 in 3 chance that Mr X chairs a committee meeting, and 20 per cent of the meetings are both chaired by Mr X and last for more than 1 hour. If Mr X chairs the committee meeting tonight, what are the odds it will last for more than 1 hour?

10 In a work-shop there are three machines producing a particular article. An inspector is equally likely to choose to sample articles from the second and third machines, and twice as likely to choose the first machine as the second. Twelve per cent of the articles produced by the first machine, 10 per cent of those produced by the second, and 15 per cent of those produced by the third, are defective.

What is the probability that the article the inspector samples:

i is from the first machine,

ii comes from the first machine given that the article is defective,

iii is defective?

5
Probability distributions

In chapter 3 we discussed how to summarize the location and dispersion of a frequency distribution and describe its shape. It might also be useful if we could find a function which would give us the frequencies $f(X)$ from the values X. This is a complicated matter, for when dealing with real data the number of possible frequency distributions we could encounter is limitless. Tackling the problem from the data end could prove to be unfruitful and we might do better to consider theoretical distributions, perhaps starting from functions, and see whether these are any use in practical situations. Four standard distributions are described in this chapter.

Discrete distributions

When dealing with frequency distributions, distinction has to be made between discrete variables and continuous variables. Observations of a discrete variable are sometimes left ungrouped and sometimes grouped into classes. Observations of a continuous variable are grouped into classes. Theoretically, observations of a discrete variable can always be left ungrouped. The practical reason for not doing so with actual data in some cases is that scarcity of observations would mean many zero and very small frequencies, thus leaving data ungrouped, while minimizing the summarizing of the data could disguise any pattern in the distribution. If the reader considers leaving the data of chapter 2, question 7, ungrouped he will appreciate these points better. Having said that, since in theoretical analysis this difficulty does not occur, discrete variables are assumed to be in ungrouped distributions in this chapter. As seen in chapter 2, a discrete frequency distribution consists of a set of values X together with their frequencies $f(X)$ (or relative frequencies $f(X)/\Sigma f(X)$).

In a similar way, remembering that the relative frequency of X is used to estimate the probability of the value X (see chapter 4) a (theoretical) discrete probability distribution consists of a set of values X together with the probabilities $p(X)$ that values X occur. As $p(X)$ is a probability it cannot be negative, though it might be zero. Since the occurrences of the values X are exhaustive and mutually exclusive events, $\Sigma p(X) = 1$ (just as the sum of relative frequencies for a distribution of data is one).

Thus the *discrete density function* $p(X)$ is defined by

$$\text{(i)}\quad p(X) \geq 0$$
$$\text{(ii)}\quad \Sigma p(X) = 1$$

where $p(X)$ is the probability that X occurs. Checking that points (i) and (ii) hold enables us to say whether or not we have a density function.

For example, suppose X denotes the number obtained when an ordinary six-faced fair die is tossed. Then we know that $p(X) = 1/6$ for every X, so $p(X) \geq 0$ for every X and $\Sigma p(X) = 1$. The set of X with corresponding $p(X)$ is a discrete probability distribution.

As a second example suppose that we are told that X takes value -1 with probability $p(-1) = 0{\cdot}20$, value 0 with probability $p(0) = 0{\cdot}50$, and can also take the value $+1$. Then, for $p(X)$ to be a density function we need

$$p(-1)+p(0)+p(+1) = 1 \text{ with } p(+1) \geq 0$$

Substituting values for $p(-1)$ and $p(0)$ gives

$$\begin{aligned} p(+1) &= 1-0{\cdot}20-0{\cdot}50 \\ &= 0{\cdot}30 \text{ which is } \geq 0 \end{aligned}$$

The probability distribution is

X	$p(X)$
-1	0·20
0	0·50
1	0·30

The diagrammatic representation of a discrete probability distribution is a line chart with the lengths of the lines drawn proportional to probabilities $p(X)$ (so that the sum of the lengths of the lines is one). Compare this with the diagrammatic representation of a discrete frequency distribution by a line chart.

The arithmetic mean of a discrete probability distribution is also referred to as the expected value of X (written $E(X)$) and is defined to be

$$\mu \text{ (or } E(X)) = \Sigma X\, p(X)$$

Notice the similarity of this to the arithmetic mean $\bar{X}$ of a frequency distribution which is

$$\bar{X} = \Sigma X \frac{f(X)}{n}$$

where $\frac{f(X)}{n}$ is the relative frequency of X.

The variance of a discrete probability distribution is defined to be

$$\sigma^2 = \Sigma(X-\mu)^2\, p(X)$$

Compare this with the variance

$$S^2 = \Sigma(X-\bar{X})^2 \frac{f(X)}{n}$$

of a frequency distribution.

All statements relating to changes in the arithmetic mean and the variance caused by changing the origin or the scale of measurement of X hold as for frequency distributions with $\bar{X}$ replaced by μ, 'variance' $= S^2$ by σ^2, and relative frequency by $p(X)$. The computational form of the formula for the variance also holds, and, as with sample data, should be used in preference to the form given as a definition except in a few cases. The computational form is now

$$\sigma^2 = \Sigma X^2\, p(X) - \mu^2$$

If we cumulate probabilities (as we cumulate frequencies in frequency distributions) we obtain the *cumulative distribution function*

$$\begin{aligned} F(a) &= \text{prob (value} \leq a) \\ &= \sum_{X \leq a} p(X) \end{aligned}$$

Continuous distributions

The change from frequency distributions of continuous variables to continuous probability distributions is not quite so easy as the change in the discrete case. We pointed out in chapter 2 that with observations on continuous variables, the same value rarely occurs twice although values may be very close, and we also remarked that the value recorded could well be an approximation, because measuring apparatus cannot give a measurement to an infinite number of decimal places. We deal with this problem in frequency distributions by grouping values into classes and finding the frequency $f(X)$ of the class whose mid-point is X. Similarly, when in theory we come to continuous probability distributions it is useful to consider values grouped into classes, in this case classes of infinitesimal width. Scarcity of observations is no worry in theory.

In continuous probability distributions we denote values by X. The number of possible values which could occur is infinite, so the probability of any value actually occurring is said to be zero. In order to understand how we measure the probability of getting a value between two specified values, that is in a 'class', consider the pictorial representation of a grouped frequency distribution by a histogram (p. 20). In the histogram the areas of blocks are proportional to the frequencies or relative frequencies of the classes of the distribution. If we think of the areas of the blocks as being equal to the relative frequencies, which are estimates of the probabilities of getting values in the different classes, the total area of the histogram is then one.

Now suppose that the total frequency is increased so that it is feasible to group the data into a larger number of narrower classes. In the histogram, the areas of blocks can still be thought of as being equal to the relative frequencies of the classes, and the total area of the histogram is still one. In theory we can repeat this process indefinitely. In the limit we should have an infinite number of classes, each of infinitesimal width. The histogram, as defined by the tops of the blocks, would appear to be a curve. The area of each block of infinitesimal width would be thought of as the probability of getting a value in that class rather than the relative frequency of that class. The total area of the histogram would be the area under a curve and would still be one.

Consider a block of infinitesimal width centred on the value X. Suppose that the width of the block is dX, where dX is infinitesimally small, so that the base of the block extends from value $X-\frac{1}{2}dX$ to value $X+\frac{1}{2}dX$. (Mathematically, as in calculus, we should first consider the width to be of small finite width δX and then take limits as δX tends to zero.) Suppose that the equation of the curve is $p(X)$ so that the height of the block is $p(X)$ at point X. Then the area of the block is $p(X)\,dX$ and $p(X)\,dX$ is said to be the probability of getting a value between $X-\frac{1}{2}dX$ and $X+\frac{1}{2}dX$. The probability of getting a value between values a and b is found by summing together the areas of all the blocks of infinitesimal width between a and b (using the process of integration). This gives the area under the curve $p(X)$ between a and b, written mathematically as $\int_a^b p(X)\,dX$ (read as 'the integral from a to b of $p(X)\,dX$'). (Knowledge of the technique of integration is not required in this book.)

The *continuous density function* $p(X)$ is defined by

$$\text{(i)} \quad p(X) \geq 0$$

(ii) the area under the curve $p(X)$ is one (written mathematically

$$\int_{-\infty}^{\infty} p(X)\,dX = 1\,)$$

where $p(X)\,dX$ is the probability of getting a value between $X-\frac{1}{2}dX$ and $X+\frac{1}{2}dX$. Checking that points (i) and (ii) hold, enables us to say whether or not we have a density function. A continuous probability distribution consists of the function $p(X)$, together with the values of X for which that $p(X)$ holds. The function $p(X)$ could be different for different ranges of values taken by X, and might be zero for some X. Thus limits $-\infty$ and $+\infty$ used with the integration sign above (and some below) are in practice replaced respectively by the lowest and highest values of X in the distribution.

The diagrammatic representation of a continuous probability distribution is the curve of the density function $p(X)$. Values X are plotted as abscissae with the corresponding values of $p(X)$ as ordinates. Areas under the curve represent probabilities, and the total area under the curve is one.

In order to find the mean μ of a continuous probability distribution we need to sum expressions 'X times the probability of getting a value between $X-\frac{1}{2}dX$ and $X+\frac{1}{2}dX$ (i.e. $p(X)\,dX$)' over all the blocks of infinitesimal width under the curve $p(X)$ (compare the expression for the mean of a discrete probability distribution). The summation is achieved by integration, so that mathematically we define

$$\mu \text{ (or } E(X)) = \int_{-\infty}^{\infty} X\,p(X)\,dX$$

Similarly, to find the variance σ^2 we sum expressions '$(X-\mu)^2$ times the probability of getting a value between $X-\frac{1}{2}dX$ and $X+\frac{1}{2}dX$' over all blocks of infinitesimal width, or, mathematically

$$\sigma^2 = \int_{-\infty}^{\infty} (X-\mu)^2\,p(X)\,dX$$

The equivalent computational form for the variance is

$$\sigma^2 = \int_{-\infty}^{\infty} X^2\,p(X)\,dX - \mu^2$$

Statements relating to changes in the arithmetic mean and the variance caused by changes in the origin and scale of X hold as in the discrete case.

The *cumulative distribution function* is

$$\begin{aligned} F(a) &= \text{prob (value} \leq a) \quad \text{(compare the discrete case)} \\ &= \text{area under } p(X) \text{ from } -\infty \text{ to } a \\ &= \int_{-\infty}^{a} p(X)\,dX \end{aligned}$$

Thus prob (value between a and b) = prob (value $\leq b$) − prob (value $\leq a$)

$= F(b) - F(a)$ which is $\int_a^b p(X)\,dX$

(Compare the use of the ogive to find frequencies from values, p. 44.)

The binomial distribution

This is an example of a discrete probability distribution. Suppose that we have an experiment which has only two outcomes. In fact any experiment can be considered to have only two outcomes, a specific outcome 'E', and all other possible outcomes 'not E'. In the binomial distribution it is convenient to refer to the two outcomes as 'success' and 'failure'. Sometimes we refer to the outcomes as being 0 and 1, where 1 corresponds to a success and 0 to a failure. We suppose that the experiment is repeated n times and we get exactly X successes (and exactly $n-X$ failures) in these n repetitions. X is a discrete variable taking values $0,1,2, ---, n$. The *binomial distribution* gives the probabilities that X takes different values.

Suppose that the probability of getting a success is p. Then the probability of getting a failure is $1-p$ and we denote this by q. We assume that p and q remain constant throughout the n repetitions of the experiment and that the n different trials of the experiment are statistically independent of one another (i.e. the outcome of any one trial does not depend on the outcome of any other trial).

Before we consider the general case of X successes in n trials let us consider a particular case, say 2 successes in 4 trials, which we could imagine to be getting 2 heads in 4 tosses of a coin, similar to examples in chapter 4. There are 16 possible outcomes of the 4 trials. Of these, 6 favour the event 2 successes (and 2 failures) in some order and since the number of trials is small we can easily list these. Denoting success by S, failure by F, we could get our 2 successes as one of the following, listing the results of each set of trials in order of occurrence.

$SSFF$ $\quad$ $SFSF$ $\quad$ $SFFS$
$FSSF$ $\quad$ $FSFS$ $\quad$ $FFSS$

In each of these 6 possible ways of getting the 2 successes, the probability of getting successes and failures in that particular order, since trials are statistically independent of one another, is p^2q^2, as is easily checked. Now the 6 possible ways of getting the 2 successes are mutually exclusive, and either we get the 2 successes in 4 trials as $SSFF$ or as $SFSF$ or as one of the other 4 sequences of results listed. Therefore we use the additive law of probabilities and find

$$\text{prob (2 successes in 4 trials)} = 6p^2q^2$$

In the case of a fair coin with S corresponding to 'head' $p = q = \frac{1}{2}$ and

$$\text{prob (2 heads in 4 tosses)} = 6 \times \tfrac{1}{4} \times \tfrac{1}{4} = \tfrac{3}{8}$$

Now consider the general case of X successes in n trials. We cannot list the ways in which we can get the X successes unless we give X and n specific values, but we can say that the probability of getting X successes and $n-X$ failures in a specified order in n repetitions of the experiment is $p^X q^{n-X}$, whatever the order of the results. The next step, referring back to the example of 2 successes in 4 trials, is to use the additive law. To do so we need to know the number of ways in which we can get exactly X successes in n trials. We can think of this as the number of ways of choosing which X trials in the n are to be successes. A general formula for this number, which is denoted by ${}^nC_X$ (read as $n\ C\ X$) can be derived fairly easily (see Kemeny *et al.*, 1957). Therefore we can say

$$\text{prob } (X \text{ successes in } n \text{ trials}) = {}^nC_X\, p^X\, q^{n-X} \qquad (X = 0,1,2,\ldots,n)$$

$$\left(\text{where } {}^nC_X = \frac{n!}{X!\,(n-X)!} \text{ and } n! \text{ is } n \text{ factorial}\right)$$

and this is the *binomial distribution*. Here n and p are the parameters of the distribution, that is, they are fixed in any one distribution in contrast to X which varies. Changing n and p in the expression above gives different binomial distributions.

In order to use the distribution we need to be able to find ${}^nC_X$ when n and X have specific values. One way of doing this is to use Pascal's triangle. The first few lines of this are reproduced below

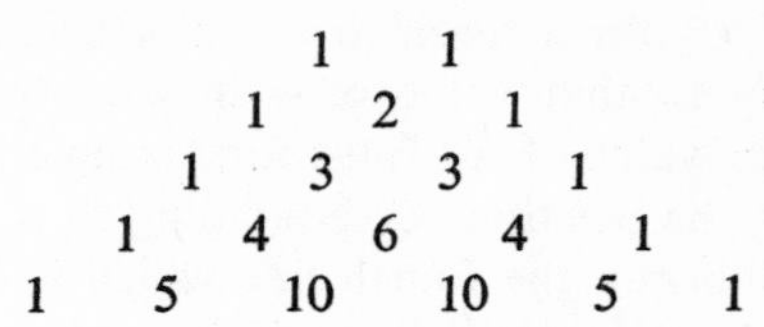

The triangle is very easy to construct. Each line starts and finishes with a 1 and the other numbers can be found from the numbers in the line above. For example, consider the fourth line. The first and last numbers are 1. The second number in the line is 4, which is the sum of the numbers 1 and 3 in the previous line and is written underneath the space between 1 and 3. The third number in the line is 6 which is the sum of 3 and 3 in the line above, and is written underneath the space between them, the fourth number is 4 which is the sum of 3 and 1 in the line above.

Now consider finding the fifth line. We know that the first and last numbers of the fifth line are 1. The other numbers are found from the fourth line, and in order are $1+4 = 5$, $4+6 = 10$, $6+4 = 10$, $4+1 = 5$. Each number is written underneath the space between the two numbers whose sum it is. Notice that the second number in each line is the number of the line, for example the second number of the second line is 2, the second number of the third line is 3.

Numbers ${}^nC_X$ are found from Pascal's triangle as follows. The lines of the triangle correspond to values of n, the first line to $n = 1$, the second to $n = 2$, the third to $n = 3$, and so on. Values of X are determined by values of n since X takes integer values between 0 and n inclusive. For example, if $n = 4$ then X takes values 0,1,2,3,4. The positions of numbers within a line of the triangle correspond to values of X, counting from left to right. Thus, in the fourth line the first number corresponds to $X = 0$, the second to $X = 1$, the third to $X = 2$, the fourth to $X = 3$, and the fifth to $X = 4$. When counting remember that X starts at 0, not at 1. The number ${}^nC_X$ for a specified n and X occurs in the line corresponding to that value of n in the position corresponding to the value of X. Thus, for example, ${}^4C_2$ is in the fourth line in the position corresponding to $X = 2$, so it is the third number in the fourth line which is 6. ${}^3C_3$ is the fourth number in the third line which is 1.

It would be impractical to use Pascal's triangle for large values of n but tables of the coefficients ${}^nC_X$ exist and could be used. In the binomial distributions discussed in this book Pascal's triangle will be satisfactory. Under certain conditions the binomial distribution can be approximated by other distributions for large values of n, particularly by the Poisson distribution (see Walpole, 1974, and Yeomans, 1968) and the normal distribution (covered later in this chapter).

As with all discrete probability distributions, the diagrammatic representation of the binomial distribution is a line chart, the exact form depending on the values of n and of p. The arithmetic mean of the binomial distribution is np and can be found directly as the sum of quantities $X\,{}^nC_X\,p^Xq^{n-X}$ over all values of $X(X = 0,1, \ldots, n)$ or in other ways, but notice that reasoning says if in n trials there is a probability p of success at each trial, we expect to get successes in proportion p of the trials, that is, we expect np successes on average. The variance of the binomial distribution is npq.

The binomial distribution, although relating to a simple situation of an experiment with only two outcomes, is very wide in its application. Frequently, in situations with several possible outcomes, we find that we are interested in just one particular outcome and how many times it occurs. In such cases it is appropriate to use the binomial distribution, provided we are able to make the necessary assumptions of a constant probability of 'success' and of statistical independence of 'trials'. For example, delivery of goods from a wholesaler to a retailer might be on any day after the order has been given (several possibilities). The retailer wants goods delivered on or before a specified day and considers all late deliveries to be useless, regardless of how many days they are late. He might wish to calculate the probability that in five orders placed with the wholesaler at least three of them will arrive on or before the day specified; that is, on time. Suppose that he knows from experience that one-third of his orders arrive on time and that deliveries are independent of one another. Then, taking arrival of an order on time as a success, the probability of a success $= p = \frac{1}{3}$ and the binomial distribution tells us

$$\text{prob (3 of 5 orders arrive on time)} = {}^5C_3 \left(\tfrac{1}{3}\right)^3 \left(\tfrac{2}{3}\right)^2 = \frac{10 \times 4}{3^5}$$

$$\text{prob (4 of 5 orders arrive on time)} = {}^5C_4 \left(\tfrac{1}{3}\right)^4 \left(\tfrac{2}{3}\right)^1 = \frac{5 \times 2}{3^5}$$

$$\text{prob (5 of 5 orders arrive on time)} = {}^5C_5 \left(\tfrac{1}{3}\right)^5 \left(\tfrac{2}{3}\right)^0 = \frac{1}{3^5}$$

The probability that at least 3 of the 5 orders are on time is the sum of these three probabilities, that is,

$$\frac{1}{3^5}(40+10+1) = \frac{51}{3^5} = \frac{17}{81}$$

Some other more obvious applications of the binomial theorem are in testing whether or not an article is defective, when the probability that a specified proportion of articles in a batch is defective might be of interest, observing whether or not a product is effective, for instance whether or not a soap powder removes stains or a drug cures complaints, and in considering conventional failure/success situations such as passing examinations.

The negative binomial distribution

The negative binomial distribution is another discrete probability distribution which is easy to derive and which is very useful in application. As with the binomial distribution we have an experiment with two outcomes which we shall call success and failure, the probability of success being p and remaining constant throughout different repetitions of the experiment, and different trials of the experiment being statistically independent of one another. We repeat the experiment until we have obtained a fixed number k of successes. Suppose that we have to repeat the experiment X times in order to achieve these k successes. Then X might be equal to $k, k+1, k+2, \ldots$. The *negative binomial distribution* gives the probabilities that X takes these different values.

Clearly, we stop repeating the experiment as soon as we have obtained k successes so the kth success occurs on the last trial, that is, the Xth trial. In the first $X-1$ trials there are $k-1$ successes (and $X-k$ failures). We can therefore say that

prob (need X trials for k successes)

= prob ($k-1$ successes in first $X-1$ trials and success on last trial)

= prob ($k-1$ successes in first $X-1$ trials) prob (success on last trial)

since trials are independent.

Referring to the binomial distribution we know that

prob ($k-1$ successes in $X-1$ trials)

$$= {}^{X-1}C_{k-1}\, p^{k-1} q^{X-k}$$

We also know that

prob (success on last trial) $= p$

$\therefore$ prob (need X trials for k successes)

$$= {}^{X-1}C_{k-1}\, p^{k-1}\, q^{X-k}\, p$$

$$= {}^{X-1}C_{k-1}\, p^{k}\, q^{X-k} \quad (X = k, k+1, k+2, \ldots)$$

and this is the *negative binomial distribution.*

The arithmetic mean of this distribution is k/p. Thus, for example, if the probability of success, p, is $\frac{1}{4}$ we expect to need $4k$ trials on average in order to get k successes, as common sense tells us.

The negative binomial distribution is applicable whenever the binomial distribution is. It is of particular use in decision making applications of the type where an important factor is how long it takes to get a certain number of successes or failures. For example, the retailer discussed previously (p. 101) might decide that if there is less than half a chance that he needs to place fewer than five orders in order to get three deliveries on time he will change to another wholesaler. Here, as before, a delivery on time is a 'success' and $p = \frac{1}{3}$. The number of successes required, k, is 3, and the retailer is interested in needing either three or four 'trials' (orders) for three successes.

In the case of three trials $X = 3$ and the negative binomial distribution tells us that

$$\text{prob (need 3 trials for 3 successes)} = {}^2C_2(\tfrac{1}{3})^3(\tfrac{2}{3})^0 = \frac{1}{27}$$

In the case of four trials $X = 4$ and the negative binomial distribution tells us that

$$\begin{aligned}\text{prob (need 4 trials for 3 successes)} &= {}^3C_2(\tfrac{1}{3})^3(\tfrac{2}{3})^1\\ &= 3\left(\frac{2}{3^3}\right)\\ &= \frac{2}{9}\end{aligned}$$

$$\therefore \text{prob (need fewer than 5 orders)} = \frac{1}{27}+\frac{2}{9}=\frac{7}{27}$$

Since 7/27 is less than ½ the retailer will decide to change to another wholesaler.

The normal distribution

The normal distribution is a continuous probability distribution. It is so called because many sets of real data have distributions which are exactly or approximately normal and because it occurs a great deal in theory so that it is 'normal' for this distribution to occur. It is a very important distribution because of its frequent occurrence. It is sometimes called the Gaussian distribution after C. F. Gauss (1777–1855) though he was not the first to discuss it.

The normal distribution is symmetrical and is described as being 'bell-shaped'. Thus, its arithmetic mean, median, and mode all coincide at the middle value of the distribution. Values taken in the distribution can be anything between $-\infty$ and $+\infty$, but as shown in Figure 5.1 the bell shape means that very few extreme negative or extreme positive values occur. The curve of the normal distribution density

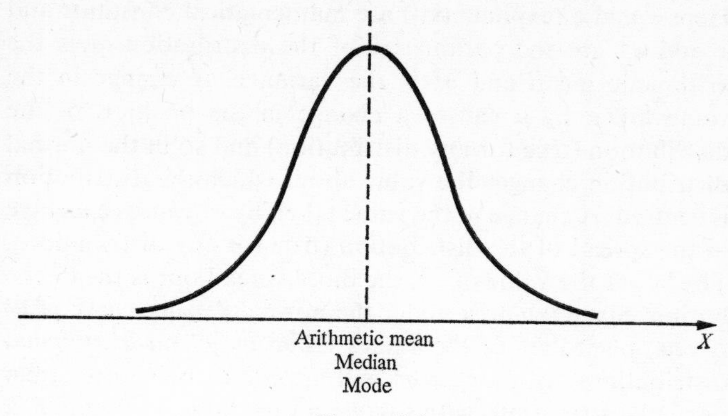

Figure 5.1

function gets closer and closer to the X axis the further the values X from the middle value.

As shown by Figure 5.1, the bulk of the values are moderately close to the arithmetic mean. If we take the two values which are the distance of one standard deviation (S.D.) away from the mean, that is, the values 'arithmetic mean minus 1 S.D.' and 'arithmetic mean plus 1 S.D.' we find that approximately 68 per cent of the distribution consists of values between these two. We say that 68 per cent of the distribution lies within one standard deviation of the mean. Moving further from the mean, we find that approximately 95 per cent of the distribution lies within two standard deviations of the mean, and slightly more than 99·7 per cent lies within three standard deviations of the mean. Thus, if we concentrate on the central set of values we are not neglecting very much of the distribution.

The density function of the normal distribution is

$$p(X) = \frac{1}{\sigma\sqrt{(2\pi)}}\, e^{-\frac{1}{2}(X-\mu)^2/\sigma^2}$$

Here π and e (exponential) are mathematical constants and μ and σ^2 are the parameters of the distribution. μ is the arithmetic mean and σ^2 is the variance. A change in the value taken by μ causes a change in the position of the distribution (true for any distribution) and so in the normal distribution changes the value about which the distribution is centred. A change in the value taken by σ^2 causes a change in the spread of the distribution (true for any distribution). The larger the value of σ^2, the more spread out is the distribution. Since the area under the normal density curve $p(X)$ is one, regardless of the values taken by μ and σ^2, normal distributions with larger σ^2 will appear to be shorter than normal distributions with smaller σ^2, as shown in Figure 5.2. A spreading of the distribution without a corresponding 'shortening' would result in a gain of area.

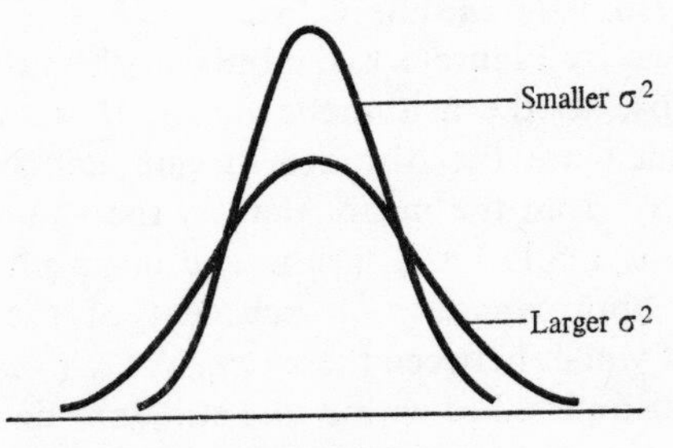

Figure 5.2

In order to find probabilities relating to normal distributions we need areas under the normal curve $p(X)$. In principle, as we know $p(X)$ we can find the areas. In practice things are not so easy. The integrations necessary to find the areas cannot be done exactly, but need to be found by using numerical approximation techniques. However, areas under the normal distribution are extensively tabulated (see Table 1 in Lindley *et al.*, 1966). We do not need a table for each pair of values of μ and σ (and there is an infinity of pairs, since μ can take any real value, σ any non-negative value) for any normal distribution can be converted into the *standard normal distribution* which has mean zero and standard deviation one. Thus, tables giving areas for the standard normal distribution can be used for any normal distribution. Notice that it is convenient to have a distribution centred on zero, and to have its standard deviation as a measure of unit. If X has a normal distribution with mean μ and standard deviation σ and we transform X to z where

$$z = \frac{X - \mu}{\sigma}$$

then z has the standard normal distribution. We say that X is $N(\mu, \sigma^2)$ and z is $N(0, 1)$. Then z is said to be a *standard normal deviate*. The standard normal distribution is pictured in Figure 5.1.

Care should be taken in using tables of the standard normal distribution as there are several different areas which appear in different tabulations, for example, some tables give the area under the curve between $-\infty$ and $+z$, some give the area between $-z$ and $+z$, some the area between 0 and $+z$, some the area between $+z$ and ∞, where z is a standard normal deviate. By symmetry, and since the total area under the curve is 1, the area under the curve between $-\infty$ and 0 is $\frac{1}{2}$ (and so is the area between 0 and ∞). Similarly, by symmetry, if the area under the curve between 0 and z is p, that is, the probability of a value between 0 and z is p, then the probability of a value between $-z$ and 0 is also p. Figure 5.3 shows how the different areas relate to one another.

To illustrate the use of tables let us suppose that X has a normal distribution with mean 3 and variance 4, that is, X is $N(3, 4)$ and we want to know the probability of getting

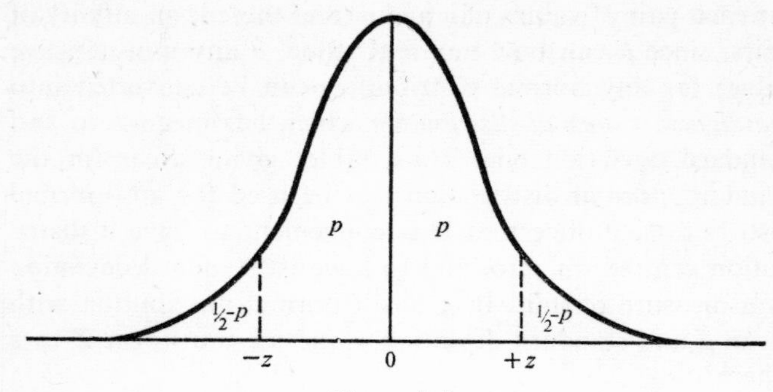

Figure 5.3

a value between 1·5 and 5·5. In order to use the tables we need to convert these values of X to standard normal deviates. Since $\mu = 3$ and $\sigma = \sqrt{4} = 2$, the standard normal deviate corresponding to the X of 1·5 is

$$z = \frac{1{\cdot}5-3}{2} = \frac{-1{\cdot}5}{2} = -0{\cdot}75$$

The standard normal deviate corresponding to the X of 5·5 is

$$z = \frac{5{\cdot}5-3}{2} = \frac{2{\cdot}5}{2} = 1{\cdot}25$$

Thus $\text{prob}\,(1{\cdot}5 < X < 5{\cdot}5) = \text{prob}\,(-0{\cdot}75 < z < 1{\cdot}25)$
$= \text{prob}\,(z < 1{\cdot}25) - \text{prob}\,(z < -0{\cdot}75)$
$= 0{\cdot}8944 - 0{\cdot}2266$ (probabilities obtained from Table 1 in Lindley *et al.*, 1966)
$= 0{\cdot}6678$

It is often helpful to draw sketch diagrams when dealing with areas under normal distributions.

The normal curve is sometimes known as the curve of errors since errors in measurement are frequently distributed normally. Measurements in themselves are sometimes distributed normally, in particular, many anthropological measurements such as heights and weights have distributions which are more or less normal.

Some of the theoretical occurrences of the normal distribution are described in chapter 6, but we mention one here since it concerns the binomial distribution discussed earlier in this chapter. Referring to the notation of the binomial distribution, if X has a binomial distribution and there are n trials with probability p of success at each, then, when n is large and p is not close to zero, X has approximately the normal distribution with mean np and variance npq. As a rule of thumb the normal approximation to the binomial is fairly good when $np > 5$ for $p \leq \frac{1}{2}$ and when $nq > 5$ for $p > \frac{1}{2}$. The approximation improves with increasing size of n. The size of the error in using the normal distribution instead of the binomial can be calculated.

To test whether or not X, given in the form of a grouped frequency distribution, has a normal distribution we can use arithmetical probability graph paper. One scale of this graph paper is uniform as in ordinary graph paper. The other scale is so constructed that if we plot the 'less than' cumulative frequency curve on the paper, using the uniform scale for the end-points of classes, the resulting graph will be a straight line, or more or less a straight line, if the distribution is normal.

The lognormal distribution

The lognormal distribution (see Aitchison *et al.*, 1969) is another example of a continuous distribution. The easy way to describe it is to say that X has a *lognormal distribution* if the logarithm of X has a normal distribution, but since the distribution has useful applications, in particular, in economics when describing distributions of incomes, it deserves some description as a distribution in its own right.

In its simplest form the lognormal distribution has two parameters corresponding to the two parameters μ and σ^2 of the related normal distribution. A variable X having a lognormal distribution takes values between 0 and ∞. The distribution has one mode and is of positive skew. The greater the value of σ^2 (the variance of the related normal distribution) the greater the skew. The curve of the distribution rises fairly steeply from zero to a maximum at the mode and then tapers off more slowly with increasing values of X as shown in Figure 5.4.

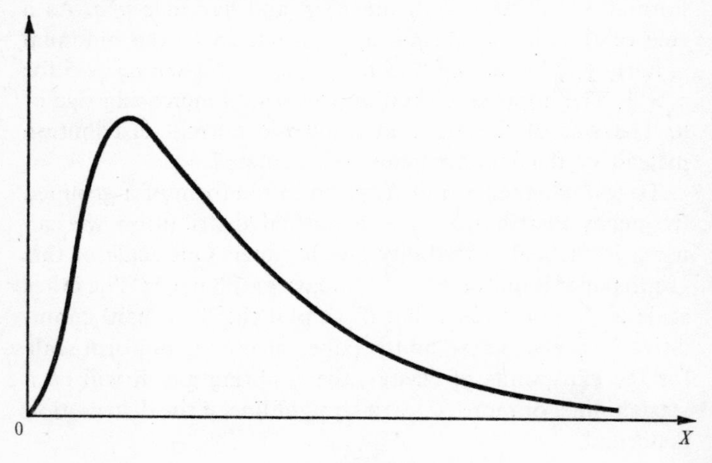

Figure 5.4

Some simple modifications enable us to make greater use of the lognormal distribution. For example, if τ is a constant such that $(X-\tau)$ has a lognormal distribution with parameters μ and σ^2, then X takes values between τ and ∞ and X is said to have a three parameter lognormal distribution with parameters μ, σ^2, and τ. The variable X now has a lower bound of τ instead of the lower bound of zero of the two parameter case. If we now consider the mirror image of the curve of this three-parameter distribution about the line $X = \tau$ we obtain a curve of negative skew for a variable taking values between $-\infty$ and τ, that is, a variable with an upper bound of τ (see Figure 5.5). This mirror image corresponds to $(\tau - X)$ having a lognormal distribution with parameters μ and σ^2. The importance of this distribution is that it is of negative skew. Notice that although the variable can now take negative values the probability of doing so is very small as the curve is close to the negative axis.

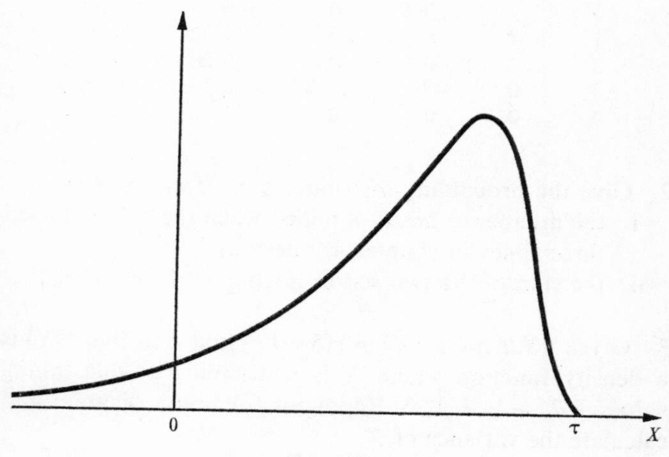

Figure 5.5

The lognormal distribution is sometimes used to approximate for discrete distributions, for example, distributions of the type 'number of households by number of resident persons' can often be well fitted by lognormal distributions. Other examples of lognormal distributions are distributions of expenditure on particular commodities, of industrial profits, and of inheritances, as well as of incomes mentioned earlier. The distribution occurs in studies of small particles and in biology where there are laws of growth which we use in statistics to generate lognormal distributions.

Exercises

1 Fill in missing values of $f(X)$ to ensure that $f(X)$ is a probability distribution.

	i	ii	iii	iv	v
X	$f(X)$	$f(X)$	$f(X)$	$f(X)$	$f(X)$
0	$\frac{1}{4}$	0·2	0	0·11	0·38
1	$\frac{3}{8}$	0·1	?	0·32	0·20
2	?	0·3	0·5	0·20	0·15
3	0	?	0·2	0·17	0·28
4	0	0	0·3	?	?

2 Give the probability distribution of X where X is:
- i the number of heads obtained when the coin is tossed three times in chapter 4, question 3
- ii the sum of the two scores in chapter 4, question 1.

3 Given a function $f(X) = c(5 - X^2)$ find c so that $f(X)$ is a density function where X is a discrete variable taking values $-2, -1, 0, 1, 2$. Represent $f(X)$ by a diagram and calculate the variance of X.

4 Represent the following binomial distributions by diagrams and label each with the mean and variance.
- i $n = 3, \quad p = 0{\cdot}3$,
- ii $n = 4, \quad p = 0{\cdot}5$.

5 There is a 3 in 4 chance that a door-bell will ring when

it is pushed. What is the probability that in 5 independent pushes of the bell it will ring no more than twice?

6 Suppose there is a constant probability of 0·6 that a dialled telephone call will reach a particular number. What is the probability that it has to be dialled (i) 4 times, (ii) 5 times, in order to reach the number 3 times? How many times does it have to be dialled on average to reach the number 3 times?

7 Given that z is a standard normal deviate find

i prob $(0{\cdot}25 < z < 1{\cdot}63)$
ii prob $(-2{\cdot}00 < z < -1{\cdot}00)$
iii prob $(-0{\cdot}43 < z < 0{\cdot}82)$
iv prob $(z < -1{\cdot}23)$
v prob $(z > 0{\cdot}27)$

8 The weights of apples in an orchard are normally distributed with mean weight 200 grams and standard deviation 25 grams:

i find the probability that an apple picked at random weighs between 150 and 230 grams
ii in what range of weights are the heaviest 20% of apples?

9 In a manufacturing process a particular set of measurements has a normal distribution with mean 57 (units) and variance 121 (units squared).

All articles with measurement larger than 79 units or smaller than 40 units have to be rejected. What proportion is rejected?

To what must the mean be changed to ensure that the proportion rejected because the measurement is too large is 2%? What total proportion is then rejected?

10 A library contains 10 000 books. The librarian finds that the probability that a book is on loan is 0·1, whatever the book. Assuming that books are borrowed independently of one another, what is the probability that the number of books on loan is:

i less than 950
ii between 980 and 1 010
iii more than 1 015?

6

Sampling distributions and their uses

In this chapter we discuss how to estimate parameters of distributions, how to measure the chance that an estimate falls in a range of values, and how to test hypotheses about the values of parameters.

Samples and populations

We need to draw a distinction between samples and populations. A *sample* is a part of a whole, and a sample of values X, possibly with frequencies, is a selection of values from a complete set of values X. The complete set with associated frequencies is a *population*. For example, the distribution of number of bedrooms in chapter 2 (p. 13) is a population, the population of 63 000 values X where X is the number of bedrooms in new houses constructed in 1970 for local authorities and new towns in England and Wales, and X takes values 1,2,3,4 or 5. There are many samples we could take from this population, such as the set of X for those houses in the 63 000 with garages, or the set of X for all of the houses situated in places with names beginning with M, or the set of X for every fifth house listed (if the 63 000 are listed in some order). All of these samples consist of values 1,2,3,4 or 5 occurring with various frequencies.

Almost any set of data can be thought of either as a sample from some underlying population or as being in itself a population. As the word population suggests people, let us take as an example data in the form of measurements X (say incomes, or ages) relating to the economists working for a large organization based in London. We could argue that the set of measurements is a population – the population of measurements X relating to the economists working for that large organization based in London. Or we could argue that the set of measurements X is a sample from one of many possible underlying populations, for example:

the population of measurements X relating to all the employees in the organization,
or the population relating to all economists in London
or the population relating to all adults in London
or the population relating to all economists working in organizations of that type
or the population relating to all economists who might have been working for the organization at present,

to name a few. In general, the context will make clear whether data form a sample or a population, and if a sample which underlying population is the appropriate one.

The populations above are all finite, but in theory we allow also for infinite populations. In practice we can achieve the effect of an infinite population by *sampling with replacement*, that is, each value taken out of the population to appear as a value in the sample is replaced in the population as soon as it has been recorded, so that the population remains the same size throughout and the same value can be re-sampled. (Consider drawing cards from an ordinary pack of 52 playing cards when each card is put back in the pack before the next is drawn.) It is possible to continue sampling for ever as it would be with an infinite population. In contrast, when *sampling without replacement* once a value has been sampled it is removed from the population and cannot be sampled again. If we sample without replacement from a finite population the population will eventually run out. A sample is said to be *random* if it is such that all samples of that size are equally likely.

Statistics and parameters

In chapter 5 we said that certain quantities are parameters of particular distributions, for example, the normal distribution has two parameters μ and σ^2. We might say that parameters are quantities belonging to populations. Thus if a population has a particular distribution, the parameters of the distribution are parameters of the population. Once we know the distribution of the population with its parameters we know everything about the population and can find other measures for it such as its mean, variance, or median, say. We may on occasions wish to refer to these also as parameters.

A measure computed from a sample is known as a *statistic* or *sampling statistic*. Thus the arithmetic mean of a sample, referred to as the sample mean and denoted by $\bar{X}$ is a statistic. The variance of a sample is a statistic. Any measure of position or spread calculated for a sample is a statistic.

If we take a random sample of size n from a population, calculate a statistic for that sample, and continue repeating this process (keeping n fixed) we build up a set of values of the statistic. The distribution of these is known as the *sampling distribution* of the statistic. Instead of continuing sampling indefinitely we can obtain the same distribution by considering all possible samples of size n if the population is finite.

For example, suppose the population consists of the five

values 0, 2, 4, 6, 8 and the statistic of interest is the arithmetic mean of random samples of size three, sampling without replacement. There are ten possible samples of size three. These are (0,2,4); (0,2,6); (0,2,8); (2,4,6); (2,4,8); (4,6,8); (0,4,6); (0,4,8); (0,6,8); and (2,6,8) with means 6/3, 8/3, 10/3, 12/3, 14/3, 18/3, 10/3, 12/3, 14/3, and 16/3 respectively.

Here the sampling distribution of the mean $\bar{X}$ is

$\bar{X}$	*Frequency*
6/3	1
8/3	1
10/3	2
12/3	2
14/3	2
16/3	1
18/3	1
	10

Estimation and properties of estimators

In many situations we have sample data but wish to make statements about some underlying population and very often this will mean using the sample data to estimate parameters for the population. For example, if we have reason to believe that heights of persons are distributed normally we might wish to estimate the mean and variance of the distribution from a random sample of heights. A statistic used as an *estimator* will take different values with different samples. Each value is an *estimate*. Some of the estimates might be exactly the value of the parameter under estimation, others might be very different from it. There is no way of telling how close any one estimate is to an unknown parameter, but we aim to choose estimators having properties which seem desirable, such as estimators giving only estimates equal to or close to the value of the parameter, for example. Thus, with the example above we should like an estimator such that if we took a second sample our estimates of the mean and variance from it would not contradict our first estimates. Similarly, we should like a procedure of estimating the mean and variance which is as satisfactory for any other normal population of heights as for the first.

Suppose that we are estimating a parameter θ by an estimator $\hat{\theta}$ calculated from a sample of size n. We are unlikely to find an estimator with all estimates equal to θ, but we should like a $\hat{\theta}$ that gives the right answer on average so that if we imagine repeating the process a large number of times some estimates err by being too large and some by being too small, but we expect the errors to cancel out. In this case $\hat{\theta}$ is said to be an *unbiased* estimator of θ. Contrast this with a biased estimator that always gives estimates of θ which are too large, say. Mathematically, $\hat{\theta}$ is an unbiased estimator of θ if the arithmetic mean of the sampling distribution of $\hat{\theta}$ is equal to θ, or in other words if the expected value of $\hat{\theta}$ is θ (written $E(\hat{\theta}) = \theta$).

We should not have much faith in an estimator of wide variability since a single estimate could then be very different from θ. We measure variability by taking the average squared deviations from $\hat{\theta}$, that is, the arithmetic mean (expected value) of $\Sigma(\hat{\theta} - \theta)^2$ and try to choose estimators with this quantity as small as possible. Comparison of this quantity for different estimators of the same θ gives a measure of *efficiency* of the estimator.

An estimator which uses all relevant information in the sample is said to be *sufficient*. For example, if we are estimating the minimum value of a population, say the minimum income earned by civil engineers, we know that this minimum income cannot be higher than the lowest income in a sample of incomes, so that all relevant information in estimating the minimum of the population is contained in one value. We cannot improve our estimation by considering any additional sample values (and would be wrong estimating the minimum by any other sample value). The minimum value in the sample is a sufficient estimator for the minimum value in the population.

An estimator which improves with increasing sample size, that is, gives estimates closer to θ as the sample size is increased, is said to be *consistent*. In practice, we need to consider also the time and cost of using different estimators in searching for a good one. It is not always possible to find a single estimator possessing all the properties thought desirable, but estimators found from the three main methods of estimation of maximum likelihood, the method of moments, and Bayes's method (see Wonnacott and Wonnacott, 1972), have good properties.

Estimation of the mean and variance of a population

Since, as we have already discussed, the arithmetic mean is important as a measure of location, and the variance as a measure of spread, and both are important as parameters of the normal distribution, estimation of the mean and variance of a population from a sample is a common problem.

We usually estimate μ, the mean of a population, by $\bar{X}$, the mean of a sample. $\bar{X}$ is an unbiased estimator of μ, as can be verified by referring to the example above, giving the sampling distribution of a mean. Here μ is $(0+2+4+6+8)/5=4$, and the mean of the sampling distribution is 12/3 (by symmetry) that is, is also 4. The variability of $\bar{X}$ is small compared with other estimators of μ, and $\bar{X}$ is a consistent estimator of μ. Using S^2, the variance of a sample, to estimate σ^2, the variance of a population, is not so satisfactory since S^2 is a biased estimator of σ^2. The unbiased estimator of σ^2 found from a sample of size n is the quantity

$$s^2 = \frac{\Sigma(X-\bar{X})^2}{n-1}$$

If we are calculating the variance of data to learn something about an underlying population, the exercise is really estimation and we should find s^2, but if the data are thought to constitute a population in their own right, S^2 is the appropriate measure. In any case, notice that if n is large s^2 and S^2 are close in value.

If we calculate the values of $\bar{X}$ and s^2 from a single sample we obtain *point estimates* of μ and σ^2 respectively. Now imagine what would happen if we took several samples and obtained a set of unbiased estimates of a parameter from them. Consideration of the sampling distribution of the estimator tells us that estimates will tend to cluster around the value of the parameter, that is, the mean of the distribution, giving a range of possible values for the parameter, and we should be more certain it was in this range than that it was equal to any one of the estimates. In fact there is no need to take several samples for we can find such a range from a single estimate and knowledge of the sampling distribution. We illustrate the method by considering estimation of the mean. An *interval estimate* of a parameter estimates the parameter as being in a range of values and gives the probability that it is in that range.

Interval estimation of the mean (confidence limits)

Suppose that we have a population with mean μ and variance σ^2 and $\bar{X}$ is the arithmetic mean of a random sample of size n. The *central limit theorem* says that the distribution of $\bar{X}$ tends to normal with mean μ, variance σ^2/n, as n tends to infinity, whatever the density of the population (discrete or continuous). This means that we can say that $\bar{X}$ is approximately normally distributed with mean μ, variance σ^2/n, for large n, and $n > 50$ is large enough for the approximation to hold fairly well. The nearer the population is to a normal population, the lower the value of n needed for the approximation to be good. In fact if the population is (exactly) normal, the distribution of $\bar{X}$ is (exactly) normal. This theorem is very important in statistics. If sampling is without replacement from a small finite population of size N the variance of $\bar{X}$ is

$$\frac{\sigma^2}{n}\left(\frac{N-n}{N-1}\right).$$

Now we know that 95% of standard normal deviates z lie between the values $-1{\cdot}96$ and $+1{\cdot}96$, as can be checked with tables of the standard normal distribution, that is,

$$\text{prob}\,(-1{\cdot}96 \leq z \leq +1{\cdot}96) = 0{\cdot}95$$

When $\bar{X}$ can be taken to have a normal distribution with mean μ, variance σ^2/n, the standard normal deviate is

$$z = \frac{\bar{X}-\mu}{\sigma/\sqrt{n}}$$

and the statement reads

$$\text{prob}\left(-1{\cdot}96 \leq \frac{\bar{X}-\mu}{\sigma/\sqrt{n}} \leq +\,1.96\right) = 0{\cdot}95$$

The inequality can be re-written to give the statement

$$\text{prob}\left(\bar{X}-1{\cdot}96\frac{\sigma}{\sqrt{n}} \leq \mu \leq \bar{X}+1{\cdot}96\frac{\sigma}{\sqrt{n}}\right) = 0{\cdot}95$$

and this holds for 95% of the values $\dfrac{\bar{X}-\mu}{\sigma/\sqrt{n}}$, that is for

95% of the values $\bar{X}$, since μ and σ are fixed. Thus, given a single sample value $\bar{X}_0$ (X nought bar) we say that we have 95% confidence that μ lies between the limits $\bar{X}_0 - 1{\cdot}96\frac{\sigma}{\sqrt{n}}$ and $\bar{X}_0 + 1{\cdot}96\frac{\sigma}{\sqrt{n}}$. These limits are called the 95% confidence limits for μ and bound an interval estimate of μ, the 95% confidence interval for μ.

For example, suppose that X is the time spent waiting to be served at a newspaper stall and we know that X has a normal distribution with mean μ minutes and variance $\sigma^2 = 0{\cdot}49$ minutes squared (so $\sigma = 0{\cdot}7$ minutes). In a sample of $n = 100$ persons, the average time spent queueing is $\bar{X}_0 = 3{\cdot}1$ minutes. Then the 95 per cent confidence limits for μ are

$$3{\cdot}1 \pm 1{\cdot}96\frac{\sqrt{0{\cdot}49}}{\sqrt{100}}$$

$$= 3{\cdot}1 \pm 0{\cdot}14$$

$$= 2{\cdot}96 \text{ and } 3{\cdot}24 \text{ (correct to 2nd decimal place).}$$

We have 95 per cent confidence that μ lies between 2·96 and 3·24 minutes.

It is possible to find other 95 per cent confidence limits for μ centred on other values, but the interval centred on $\bar{X}$ is the shortest and so gives more information about the value of μ. The choice of 95 per cent here is arbitrary. Any level of confidence c per cent can be taken and, referring to normal distribution tables, we take multipliers z_c so that

$$\text{prob}\,(-z_c \leq z \leq +z_c) = \frac{c}{100}$$

giving $\bar{X}_0 \pm z_c\frac{\sigma}{\sqrt{n}}$ as the c per cent confidence limits for μ.

This is illustrated in Figure 6.1. The most usual levels of confidence are 90 per cent, 95 per cent, and 99 per cent.

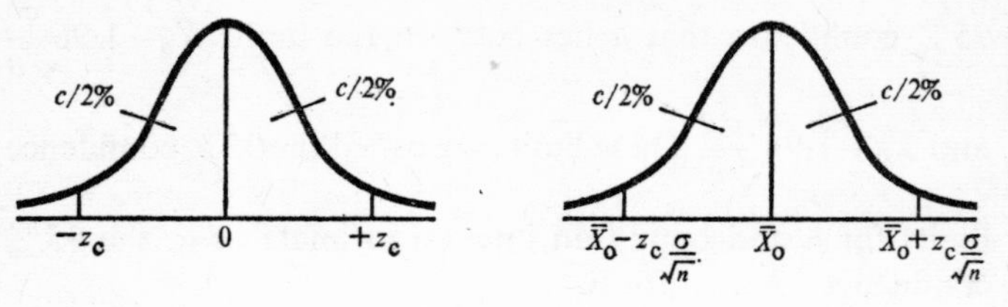

Figure 6.1

For example, taking $c\% = 98$ per cent

$$\text{prob}\,(-2{\cdot}33 \leq z \leq +2{\cdot}33) = \frac{98}{100}$$

so that 98 per cent confidence limits for μ in the problem above about the newspaper stall are

$$3{\cdot}1 \pm \frac{2{\cdot}33\sqrt{0{\cdot}49}}{\sqrt{100}}$$

$$= 2{\cdot}94 \text{ and } 3{\cdot}26 \text{ (correct to 2nd decimal place).}$$

Notice that the confidence interval becomes wider as the confidence level is increased. Put another way, if we want to narrow down our interval estimate of μ we are less certain that μ falls in the interval. Notice, however, that for fixed c per cent confidence the width of the confidence interval is $2z_c\,\sigma/\sqrt{n}$, so we can narrow the interval estimate of μ for a fixed level of confidence by increasing the sample size n.

These confidence limits for μ which we use when estimating an unknown μ, involve σ. Frequently, σ too is unknown. In this case, provided n is large, we can estimate σ by s or S (s and S are more or less equal for large n) to obtain the confidence limits. For example, in the newspaper stall problem above, if we know that X has a normal distribution and do not know either the mean μ or variance σ^2, but for the sample of $n = 100$ persons know $\bar{X}_0 = 3{\cdot}1$ minutes (as before), and know also that $s = 0{\cdot}6$ minutes, then the 95 per cent confidence limits for μ are estimated to be

$$\bar{X}_0 \pm 1{\cdot}96 \frac{s}{\sqrt{n}}$$

$$= 3{\cdot}1 \pm \frac{(1{\cdot}96)\,(0{\cdot}6)}{10}$$

$$= 2{\cdot}98 \text{ and } 3{\cdot}22 \text{ (correct to 2nd decimal place).}$$

If n is small, σ is unknown, and the population is normal, we can use a distribution called the t distribution to find an interval estimate of μ (see Bugg *et al.*, 1968; Hoel, 1971; Walpole, 1974; Wonnacott and Wonnacott, 1972; Yeomans, 1968).

Other confidence limits

Confidence limits for other parameters estimated by statistics having normal distributions can be found in a similar manner to confidence limits for the mean by starting with a probability statement involving the standard normal deviate z and rewriting it to get the parameter in the middle of the inequality. All c per cent confidence limits based on the normal distribution are of the form statistic $\pm z_c$ times standard deviation
where

$$\text{prob}\,(-z_c \leq z \leq +z_c) = c/100.$$

In particular we can obtain confidence limits for a difference in means and for a proportion.

Suppose first that $\bar{X}_1$ is the mean of a random sample of size n_1 taken from a population with mean μ_1 and variance $\sigma_1{}^2$ and that $\bar{X}_2$ is the mean of a random sample of size n_2 taken independently from a population of mean μ_2 and variance $\sigma_2{}^2$. Then, provided $\bar{X}_1$ and $\bar{X}_2$ are normal, or approximately normal, the distribution of $(\bar{X}_1 - \bar{X}_2)$ (statistic) is normal, or approximately so, with mean $(\mu_1 - \mu_2)$ (parameter) and variance $\left(\frac{\sigma_1^2}{n_1} + \frac{\sigma_2^2}{n_2}\right)$. If σ_1^2 (or σ_2^2) is unknown it can be estimated by the variance of the corresponding sample, provided n_1 (or n_2) is large. Thus, if $\bar{X}_{10}$ (X 'one nought' bar) and $\bar{X}_{20}$ are the means observed in the two samples, the $c\%$ confidence interval for $(\mu_1 - \mu_2)$ is

$$(\bar{X}_{10} - \bar{X}_{20}) \pm z_c \sqrt{\left(\frac{\sigma_1^2}{n_1} + \frac{\sigma_2^2}{n_2}\right)}$$

This distribution of the difference of means is useful in sample survey studies. For example, if we were interested in the difference between mean journey times to work in two regions, we could take samples in each region, and from the differences in mean journey times in the samples obtain an interval estimate of the population difference in means, together with the probability that the difference was in this range. Thus, if we have a sample of $n_1 = 500$ persons from region 1 with a mean journey time of $\bar{X}_{10} = 50$ minutes, and a sample of $n_2 = 650$ persons from region 2 with a mean journey time of $\bar{X}_{20} = 31$ minutes, and if the variances of the journey times in the two samples are 7 minutes and 4 minutes respectively then the 90% confidence limits for the population difference in mean journey times, $(\mu_1 - \mu_2)$, are

$$(50 - 31) \pm 1{\cdot}64 \sqrt{\left(\frac{7}{500} + \frac{4}{650}\right)}$$

$$= 18{\cdot}77 \text{ and } 19{\cdot}23 \text{ (using logs)}$$

Suppose next that there are X successes in a sample of size n, so that the proportion of successes in the sample is $P = X/n$. If the proportion of successes in the population is p, then P is an unbiased point estimator of p (this justifies estimating probabilities by relative frequencies). The sampling distribution of P is normal if the population is normal; approximately normal if n is large (say > 50) and either the population is infinite, or, if finite, sampling is with replacement. In each case the normal distribution has mean p and variance pq/n where $q = 1-p$. Notice that the standard normal deviate is

$$z = \frac{P-p}{\sqrt{(pq/n)}} = \frac{X-np}{\sqrt{(npq)}}$$

so that this is the normal approximation to the binomial distribution in another form (see p. 109). Since the variance pq/n is unknown when we are estimating p, we estimate the variance by $P(1-P)/n$ and so the c per cent confidence limits obtained for p are $P_0 \pm z_c\sqrt{[P_0(1-P_0)/n]}$ where P_0 is the proportion of successes observed in the sample.

This, too, is of use in survey studies, for example in obtaining interval estimates of the proportion of voters favouring a particular political party, or the proportion of shoppers who will buy a new brand of margarine. For example, suppose that in a sample of 300 persons, 100 are left-wing in their attitudes, so P_0, the proportion in the sample with left-wing attitudes is 100/300. If p is the proportion in the population who are left-wing in attitude the 95 per cent confidence limits for p are

$$\frac{100}{300} \pm 1{\cdot}96\sqrt{\left(\frac{100}{300}\right)\left(\frac{200}{300}\right)\Big/300}$$
$$= 0{\cdot}280 \text{ and } 0{\cdot}386 \text{ (using logs)}$$

Hypothesis testing

We can also reason from sampling distributions to obtain theory enabling us to test hypotheses concerning values of parameters in distributions. Suppose for example that we know that a population is normal and has known variance σ_0^2, say that weights of packets of biscuits are normally distributed with variance 25 square grams. We formulate a hypothesis that the mean of the population is μ_0, say that the mean weight of packets of biscuits is 227 grams. Then if $\bar{X}$ is the mean of a sample of size n the sampling distribution of $\bar{X}$ is normal with variance σ_0^2/n and under hypothesis, mean μ_0. Most of the values of $\bar{X}$ are fairly close to μ_0. In particular 95 per cent of the values lie in the range $\mu_0 - 1{\cdot}96(\sigma_0/\sqrt{n})$ to $\mu_0 + 1{\cdot}96(\sigma_0/\sqrt{n})$. Therefore if we take a random sample of size n from the population we expect its mean $\bar{X}_0$ to lie in the range $\mu_0 - 1{\cdot}96\ (\sigma_0/\sqrt{n})$ to $\mu_0 + 1{\cdot}96\ (\sigma_0/\sqrt{n})$ (in fact are 95 per cent sure that it will) and would be surprised if it lay outside this range. If the test statistic $\bar{X}_0$ lies inside the range, the hypothesis $\mu = \mu_0$ seems reasonable and we agree to accept it. Otherwise we reject it and conclude $\mu \neq \mu_0$.

Notice, however, that we could be wrong in two ways in using this test procedure. First, our sample value $\bar{X}_0$ might lie outside $\mu_0 \pm 1{\cdot}96\ (\sigma_0/\sqrt{n})$ since 5 per cent of values $\bar{X}$ do. The choice of μ_0 under hypothesis is correct, but we reject the hypothesis when we should accept it, since the sample value is extreme. This is called a Type I error and the probability of making it is the probability of obtaining an extreme value (in this case 5 per cent) shown as the area with vertical shading in Figure 6.2.

The second way in which we could be wrong is that we could obtain a value $\bar{X}_0$ inside the range $\mu_0 \pm 1{\cdot}96\ (\sigma_0/\sqrt{n})$ although the value of μ is not μ_0 but μ_0' say. In this case we accept the hypothesis that $\mu = \mu_0$ when we should reject it. This is called a Type II error. We are comparing $\bar{X}_0$ with the wrong sampling distribution. The probability of making a Type II error is the probability of getting a value from the normal distribution with parameters μ_0' and σ_0^2/n in the range $\mu_0 \pm 1{\cdot}96\ (\sigma_0/\sqrt{n})$. This depends on the value μ_0' and is the area with horizontal shading in Figure 6.2. Since μ is not known, μ_0' is any possible value of μ other than the value μ_0 under hypothesis, so there might be several probabilities of Type II error.

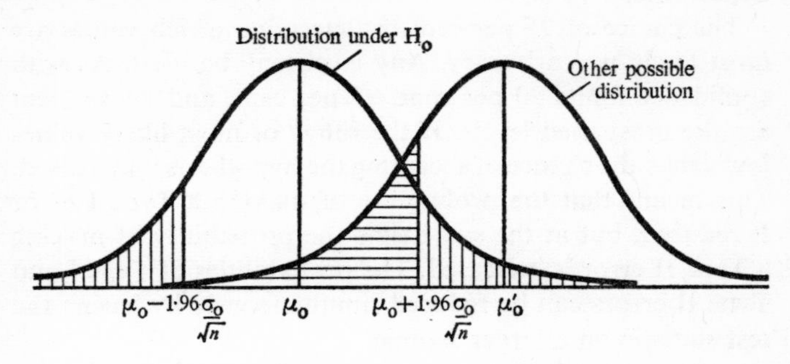

Figure 6.2 (not to scale)

A similar procedure for testing can be used with other sampling distributions. For example, whenever the sampling distribution of $\bar{X}$ can be taken to be normal we can test the hypothesis that $\mu = \mu_0$, if necessary estimating an unknown σ_0 from the sample if n is large. (If σ_0 is unknown and n is small, the t distribution has to be used.) In addition, when the sampling distribution of the difference between two sample means can be taken to be normal we can use this to test the hypothesis that means of two populations are equal, if necessary estimating variances by variances of samples. For example, we might want to test a hypothesis that the mean weight of packets of biscuits made by factory 1 is the same as the mean weight of packets made by factory 2. When the sampling distribution of a proportion can be taken to be normal we can use this to test the hypothesis that the population proportion is p_0. For example we might want to test a hypothesis that the proportion of chocolate biscuits in mixed packets is 0·25. Note that the normal distribution to use is the one with mean p_0 and variance $p_0(1-p_0)/n$ (contrast this with finding confidence limits for p where the variance had to be estimated using the sample proportion). Other distributions can also be used in testing hypotheses.

The choice of 95 per cent to determine which values are most likely was arbitrary. Any level may be used. As with confidence limits 90 per cent, 95 per cent, and 99 per cent are the most used levels. If the range of most likely values is widened the chance of accepting the hypothesis is increased. This means that the probability of making a Type I error is reduced, but at the same time the probability of making a Type II error is increased. The probabilities of Type I and Type II errors can be reduced simultaneously by basing the test statistic on a larger sample.

Terminology of tests

The hypothesis we are able to test is called the *null hypothesis* and is usually denoted by H_0 (H nought). It is sometimes formulated with the hope that it will be rejected. It involves equality, for example, we have considered testing null hypotheses $\mu = \mu_0$, $\mu_1 = \mu_2$, and $p = p_0$. Tested against the null hypothesis is an *alternative hypothesis* or research hypothesis, usually denoted by H_1 (H one). Acceptance of H_0 implies rejection of H_1, and vice versa, and H_1 might be the hypothesis of interest. Possible forms of H_1 for the null hypothesis $\mu = \mu_0$ are $\mu \neq \mu_0$, $\mu > \mu_0$ (of particular use when testing whether a change in a production process has increased the mean yield), and $\mu < \mu_0$ (of use when testing whether changes have reduced the size of a systematic error).

A *Type I error* is rejection of the null hypothesis when it is true and should be accepted. The probability of making a Type I error is called the *level of significance* of the test and is usually denoted by α. A *Type II error* is acceptance of the null hypothesis when it is false and should be rejected. The probability of making a Type II error is usually denoted by β and $1-\beta$ is known as the *power* of the test. The situation is shown in Table 6.1. The possible values under the alternative hypothesis determine the values of β and hence power is a function of these possible values.

Table 6.1

	Accept H_0	*Reject H_0*
H_0 true	Correct decision prob $= 1-\alpha$	Type I error prob $= \alpha =$ level of significance
H_0 false	Type II error prob $= \beta$	Correct decision prob $= 1-\beta =$ power

If H_1 is of the form 'parameter $\neq$ value under H_0' the test is said to be *two-sided* since all extreme values of the test statistic cause rejection of H_0. If H_1 is of the form 'parameter $>$ value under H_0' or of the form 'parameter $<$ value under H_0' the test is said to be *one-sided*, since in the first case rejection of H_0 comes only from extreme large values of the test statistic, and in the second rejection of H_0 comes only from extreme small values. This is illustrated in Figure 6.3.

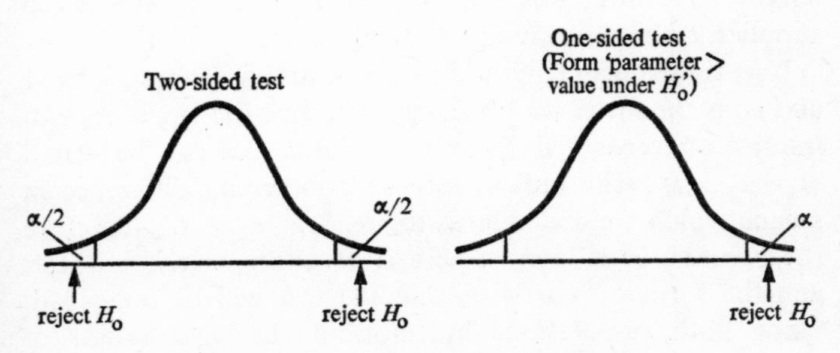

Figure 6.3

Notice that in a one-sided test one value determines the range of most likely values – the region of acceptance of H_0 – whereas in a two-sided test two values are needed. The original discussion about the test procedure was in terms of a two-sided test. The values bounding the region of acceptance (and rejection) are called *critical values.* These can be found for the sampling distribution, or the test statistic can be converted to a standard form, for example the standard normal distribution, and critical values found in terms of this.

Procedure in tests

Tests of significance need to be set out clearly and the level of significance, null and alternative hypotheses, must be clearly stated. The example below serves as a model illustrating the procedure.

Suppose that the distribution of lengths of tubing sold by each of two suppliers is normal, and that the variance of lengths of tubing from supplier 1 is $\sigma_1{}^2 = 19$ square centimetres, and from supplier 2 is $\sigma_2{}^2 = 15$ square centimetres. The hypothesis is made that the mean length of tubing from supplier 1 is greater than the mean length from supplier 2. A sample of 400 pieces of tubing from supplier 1 has mean length 72 centimetres and a sample of 200 pieces from supplier 2 has mean length 71 centimetres.

If μ_1 is the mean length of pieces of tubing from supplier 1 and μ_2 is the mean length from supplier 2 the hypothesis of interest concerns a difference in means and can be stated '$\mu_1 - \mu_2 > 0$'. The null hypothesis concerning difference in means which we are able to test is '$\mu_1 - \mu_2 = 0$', therefore the hypothesis of interest here is an alternative hypothesis and the form of it tells us that the test will be one-sided. Since both populations are normal the distribution of $\bar{X}_1 - \bar{X}_2$ is normal with mean $\mu_1 - \mu_2$ and variance

$\frac{19}{400}+\frac{15}{200}=\frac{49}{400}$ square centimetres where $\bar{X}_1$ is the mean of samples of size 400 from supplier 1, and $\bar{X}_2$ is the mean of samples of size 200 from supplier 2. We therefore set out the test as follows, having defined μ_1, μ_2, $\bar{X}_1$, $\bar{X}_2$ as above.

To test $H_0 : \mu_1 - \mu_2 = 0$

against $H_1 : \mu_1 - \mu_2 > 0$ (one-sided test)

$\bar{X}_1 - \bar{X}_2$ is $N(\mu_1 - \mu_2, 49/400)$ i.e. $N(0, 49/400)$ under H_0.

Set level of significance $\alpha = 0{\cdot}05$ (choice arbitrary here).
Sample evidence is $\bar{X}_1 - \bar{X}_2 = 72 - 71 = 1$ centimetre.
Converting sample evidence to standard normal deviate

$$z = \frac{(\bar{X}_1 - \bar{X}_2) - (\mu_1 - \mu_2)}{\sqrt{\left(\frac{\sigma_1{}^2}{n_1} + \frac{\sigma_2{}^2}{n_2}\right)}}$$

$$= \frac{1}{\sqrt{(49/400)}}$$

$$= 20/7$$

$$= 2{\cdot}86 \text{ correct to 2nd decimal place}$$

Critical value of z for $0{\cdot}05$ level of significance is $1{\cdot}64$
∴ sample value is in region of rejection for H_0.

∴ Accept hypothesis $H_1 : \mu_1 - \mu_2 > 0$, i.e. mean length from supplier 1 is greater than mean length from supplier 2 at 5 per cent level of significance.

Note that instead of converting to z we could find

$$(\mu_1 - \mu_2) + 1{\cdot}64\sqrt{\left(\frac{\sigma_1{}^2}{n_1} + \frac{\sigma_2{}^2}{n_2}\right)}$$

and compare the value of $\bar{X}_1 - \bar{X}_2$ with it. The conclusion will be the same.

Method of probability values

The method of probability values leaves the choice of level of significance open so that the reader may choose the level he thinks is appropriate. It is a useful way of presenting tests, as all the evidence is given and not hidden under the rejection or acceptance of a hypothesis. The probability value (prob value) is the probability of getting a sample result as, or more extreme than, the one observed. Thus, in the previous example, testing $H_0: \mu_1 - \mu_2 = 0$ against $H_1: \mu_1 - \mu_2 > 0$, the prob value is the probability of a standard normal deviate > 2·86, i.e. is 0·0021. Since this is less than 0·05, H_0 is rejected at the 0·05 level. On the other hand since 0·0021 > 0·001, H_0 is accepted at the 0·001 level, and so on. We say that the sample z is significant at the 0·0021 level.

If, with the data on lengths, we test $H_0: \mu_1 - \mu_2 = 0$ against $H_1: \mu_1 - \mu_2 \neq 0$, which is a two-sided test, the probability value is the probability of getting standard normal deviates greater than 2·86 or less than —2·86, which is 0·0042.

Exercises

1 Give some possible underlying populations for each of the following:

- i the heights of oak trees in Hyde Park
- ii the members of the British ABYZ Society in the current year
- iii the salaries of nursery school teachers ten years ago.

2 A population consists of four values 1, 2, 3, 4. Take samples of size two (*a*) without replacement, (*b*) with replacement, and in each case find the sampling distribution of the mean and verify that the sample mean is an unbiased estimator of the population mean.

3 i Find 92 per cent confidence limits for the mean of a population of standard deviation 42 if a sample of 5 000 values has a mean of 735.

ii Find 94 per cent confidence limits for the difference in means of populations 1 and 2 if population 1 has standard deviation 22 and a sample of 2 000 values from it has a mean of 60, population 2 has standard deviation 32 and a sample of 4 000 values from it has a mean of 70.

iii Find 96 per cent confidence limits for the proportion of successes in the population if there are 542 successes in a sample of 1 000.

4 The errors in measurement of lengths by a man are distributed normally with mean 0, standard deviation 3 centimetres.

i He measures the length of a room as 5 metres. Construct a 95 per cent confidence interval for the true length.

ii He measures the length of another room four times and obtains measurements of 4·03 metres, 3·87 metres, 4·55 metres, and 4·35 metres. What is the 99 per cent confidence interval for the true length?

5 Identify Type I and Type II errors in the following situations:

i student applying for a place at College where all students with appropriate qualifications should be given places

ii doctor prescribing a drug which cures only patients with disease X.

6 Suppose that the weights of babies at birth are normally distributed with standard deviation 0·41 kilograms. Test the hypothesis that the average birth weight is 3·18 kilograms if a random sample of 100 babies has mean weight

3·33 kilograms. Use a 5 per cent level of significance.

7 In a survey on two-person households one question asked how much money had been spent on food during the previous week. In area A the average amount spent by the 100 households in the sample was £6·78 and the standard deviation of the amounts was £0·90. In area B there were 200 households in the sample, the average amount spent was £6·53, and the standard deviation was £1.20.

Is the average amount spent by two-person households in area A significantly higher than the average amount spent in area B by such households, at the 5 per cent level of significance?

8 Two engineers make successive measurements with a micrometer of the same dimension. Engineer A obtains the 8 values 500, 501, 501, 503, 498, 499, 498, and 500. Engineer B obtains the 6 values 502, 498, 503, 502, 499, and 502.

Each engineer makes assumptions about the distributions of his and of the other's measurements as follows. Engineer A assumes that both distributions are normal with variance 3. Engineer B assumes that both distributions are normal with unknown variance σ^2.

Test the hypothesis that there is no difference between the means of the distributions of engineers A and B as if you were engineer A. On what distribution would engineer B base a test of this hypothesis?

9 Let p be the proportion of the population owning a car. In a sample of 150 persons 70 own a car. Test the hypothesis $p = 0·5$ at the 10 per cent level of significance by finding critical values of p.

10 A random sample of 100 students is taken at a certain

university. There are 45 women students in this sample. Give the probability values for tests of the hypotheses:

i more than one-third of the students at the university are women

ii there are equal numbers of men and women at this university.

7
Regression

Apart from a brief discussion of tabulation by several factors of classification in chapter 2 this book so far has concentrated on one variable at a time and has not considered relationships between variables. Now clearly for any set of units, such as persons, countries, or goods, we can list several variables for which we might have measurements. Thus, for men, we might know their incomes, their expenditure on housing, their length of time in employment, and the age at which they finished full-time education. Questions concerning relations between these variables then come to mind. For example: is expenditure on housing greater, the greater the income? Does income increase with length of time in employment when the length is fairly short, but decrease with length of time at the other end of the scale when the man is approaching retirement? What is the relation between expenditure on housing and income, for a fixed age of finishing full-time education? We require a set of methods which will enable us to answer questions such as these.

Scatter diagrams

As in the case of a single variable we start by considering

how to represent data in diagram form. Suppose first that we have just two variables X and Y, and (X_i, Y_i) is the pair of values of these variables taken by the ith unit. We show this in diagram form by a scatter diagram, which is a plot of the points (X_i, Y_i) on a Cartesian co-ordinate system – rather like an ordinary graph except that we do not join the points because in general they do not fall exactly on a line or curve but are 'scattered' around. By convention, if one of the two variables seems to depend on the other it is called the Y variable and plotted on the Y axis. The variable X is then an explanatory variable. For example, since a child's height depends on its age (and not its age on its height) height would be the dependent (Y) variable. In some cases neither variable is obviously the dependent one since both are dependent on some third variable. For example, a child's height is no more dependent on its weight than its weight is on its height, but both height and weight are dependent on age. In this case either variable (height or weight) can be called the Y variable.

The scatter diagram will show the nature of the relation between X and Y, for example, it will show if in general X and Y are increasing together. Four typical two-dimensional scatter diagrams are shown in Figure 7.1, with labels describing the relation suggested by the scatter of points.

We use the word 'correlation' in talking of the relationship. Thus, we might say that there is a positive linear relation between X and Y, or there is positive linear correlation between X and Y, or X and Y are positively linearly correlated. When the points lie close to an underlying line or curve, the correlation is said to be strong, but notice that a relation can apparently be changed by changing the scales of the axes. It is sometimes useful to identify individual points in a scatter diagram by labelling them.

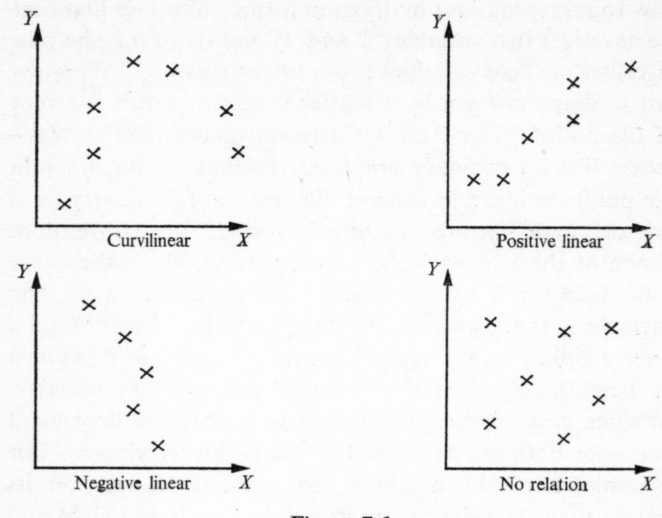

Figure 7.1

This makes it easier to investigate why certain points are off the general trend, for example.

If we have three variables we can represent the relation between them on a two-dimensional scatter diagram by plotting a scatter diagram of a pair of the variables and writing in by each point the value taken by the third variable. If one of the three variables appears to be dependent on the other two, this is the one whose values are written in on the diagram. In examining the diagram we imagine that the third variable is represented by sticks erected at right-angles to the two-dimensional diagram, the lengths of the sticks being proportional to values of the third variable. If there is a relation between the three variables the tops of these imaginary sticks describe a surface in space. It is, of course, possible to construct a model of this situation. By colouring the sticks and varying their thicknesses it is possible to indicate approximate values of fourth and fifth variables as well, but once we reach this stage algebraic descriptions of the situation are easier to understand than geometric ones.

Regression

In most cases the data plotted in scatter diagrams will be a random sample from an underlying multivariate population (the word multivariate here means that several variables are involved, whereas the populations considered in previous chapters have been univariate – involving just one variable). An observation from a multivariate population involving n variables consists of n values, a value of each variable. The complete set of observations constitutes the population. Associating with each observation the probability that it occurs, we obtain the distribution of the population.

Suppose that on the basis of sample data we decide there is a relation between variables, to keep things simple, say we decide there is a linear relation between two variables. When we think of finding the equation of the straight line describing the relation we are in a sense estimating the relation in the population as being linear, and the straight line we find using sample data is an estimate of the straight line we would find if we had complete knowledge of the population. The methods of estimating the underlying curve or surface of a relationship are known as *regression* methods, and the estimated curve (or surface) is called the regression curve (or surface).

The term 'regression' is due to Sir Francis Galton (1822–1911), although its meaning has been extended since he used it. Galton found that the sons of fathers having the same height X have heights which are on average somewhere between X and $\bar{X}$, where $\bar{X}$ is the mean height of all fathers (in the same generation), and called this 'regression towards the mean'. If Galton's observation did not hold the distribution of heights would be changing from one generation to the next. For example, if sons of fathers of height X were on average shorter than X, then the heights of sons would be less than the heights of fathers on average, implying that the human race was becoming noticeably shorter in a very small space of time.

The principle of least squares (applied to bivariate linear regression)

One of the most used methods of estimating a regression curve, or surface, is that of *least squares*. The estimators found by least squares have important theoretical properties, as we shall see to some extent later.

To illustrate the principle of least squares we discuss the simplest case, a pair of variables X, Y which are linearly related. Suppose that Y is the dependent variable. Then we want to find the equation of a straight line which we can use to predict the Y value corresponding to any specific X value in the range of X values in the sample, for the nature of the dependence of Y on X means that if we know X we know something about Y. In a sense we want the average value of Y for each X value. We say that we are regressing Y on X.

Now the general equation of a straight line is

$$Y = a+bX$$

and by varying a and b we get different lines. We want to find specific values a_0 of a, and b_0 of b, to give us the regression line. Suppose that our sample consists of n pairs of values (X_i, Y_i) $(i = 1, 2, \ldots, n)$. The value of Y on the regression line corresponding to X_i is

$$\hat{Y}_i = a_0 + b_0 X_i$$

(Notice that we can write down this algebraic expression even though we do not know a_0 and b_0 at this stage.) The deviation of the value of Y on the regression line from the actual value of Y which occurs with X_i is

$$Y_i - \hat{Y}_i = Y_i - a_0 - b_0 X_i$$

Figure 7.2 illustrates this. This deviation could be negative. The principle of *least squares* says 'choose a_0 and b_0 such that the sum of the squares of the deviations is a minimum, that is, choose a_0 and b_0 to minimize

$$\sum_{i=1}^{n} (Y_i - a_0 - b_0 X_i)^2$$

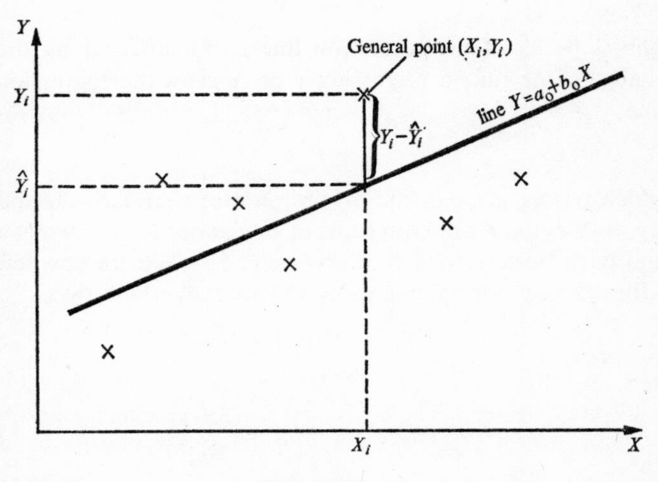

Figure 7.2

This expression can be minimized using calculus, or by an algebraic argument (both approaches are given in Wonnacott and Wonnacott, 1972). We find that we have to solve

$$\Sigma(Y_i - a_0 - b_0 X_i) = 0$$

and

$$\Sigma X_i(Y_i - a_0 - b_0 X_i) = 0$$

which are called the *normal equations*. Notice that the first of these equations says that the sum of the deviations is zero and also tells us that the regression line passes through the point $(\bar{X}, \bar{Y})$ which is the mean of the bivariate distribution (bivariate – two variables).

This means that the algebra will become easier if we change the origin to the point $(\bar{X}, \bar{Y})$ so $(\bar{X}, \bar{Y})$ becomes $(0, 0)$. This is done by transforming X to x and Y to y where

$$x = X - \bar{X}$$
$$y = Y - \bar{Y}$$

The slope b_0 of the regression line is not affected by the change of origin, so regressing y on x gives the regression line

$$y = b_0 x$$

which passes through the origin. Notice that $\Sigma x = 0$ and $\Sigma y = 0$, as these are both sums of deviations from the mean and must be zero by definition (see p. 55). We are now left with just one normal equation, the second, which says

$$\Sigma x_i (y_i - b_0 x_i) = 0$$

or

$$b_0 = \frac{\Sigma x_i y_i}{\Sigma x_i^2}$$

In terms of X and Y this is

$$b_0 = \frac{\Sigma(X_i - \bar{X})(Y_i - \bar{Y})}{\Sigma(X_i - \bar{X})^2}$$

or, re-writing it in a form suitable for computation,

$$b_0 = \frac{n\Sigma X_i Y_i - \Sigma X_i \Sigma Y_i}{n\Sigma X_i^2 - (\Sigma X_i)^2}$$

We can substitute the value of b_0 in the original first normal equation to find

$$a_0 = \bar{Y} - b_0 \bar{X}$$

b_0 is called the *coefficient of regression* of Y on X.

When fitting a regression line it is quicker and more accurate to work with the original variables X and Y and use the computational form of the formula for b_0 rather than take deviations from the means to transform to variables x and y (compare the comment made about calculating variances on p. 57). Short cuts can be made in the arithmetic by changing the origin (subtraction of constant) and/or changing the unit of measurement (division by constant) of one or both of X and Y, similar to the short cuts described in chapter 3 for calculating means and variances. These add nothing to the understanding of regression theory and there is little point in using them when calculating machines and computers are available for analysis of multivariate data and we do not describe them here.

As an example suppose that for 8 students we have the values of X = time spent studying per day and of Y = number of cigarettes smoked per day, and the scatter diagram suggests a linear relation between X and Y. It is thought that Y depends on X (though it could just be argued that X depends on Y here). The values of X and Y and the calculations needed to fit the regression line

$$Y = a_0 + b_0 X$$

are shown in Table 7.1. The values of Y^2 are also given ready for later use.

Table 7.1

X	Y	XY	X^2	Y^2
4·5	12	54·0	20·25	144
5·2	12	62·4	27·04	144
3·1	10	31·0	9·61	100
7·0	17	119·0	49·00	289
6·2	15	93·0	38·44	225
5·7	14	79·8	32·49	196
6·5	16	104·0	42·25	256
5·4	14	75·6	29·16	196
43·6	110	618·8	248·24	1 550

Substituting values in the formula for b_0 gives

$$b_0 = \frac{(8)(618{\cdot}8)-(43{\cdot}6)(110)}{(8)(248{\cdot}24)-43{\cdot}6^2}$$

$$= \frac{154{\cdot}4}{84{\cdot}96}$$

$$= 1{\cdot}817 \text{ (unrounded)}$$

Hence

$$a_0 = \frac{110}{8}-\left(\frac{154{\cdot}4}{84{\cdot}96}\right)\left(\frac{43{\cdot}6}{8}\right)$$

$$= 13{\cdot}75-9{\cdot}904$$

$$= 3{\cdot}846 \text{ (unrounded)}$$

Thus the least squares regression line for the data is

$$Y = 3{\cdot}85+1{\cdot}82X$$

We could as easily regress X on Y (or x on y) and this will make sense if X is the dependent variable (or if neither X nor Y is obviously the dependent variable). In this case the deviations in the expression we minimize are deviations between values of X on the regression line and actual values of X. The regression line is

$$X = b_1 Y + a_1$$

where

$$b_1 = \frac{\Sigma x_i y_i}{\Sigma y_i^2}$$ (with a corresponding form in terms of X_i, Y_i)

and

$$a_1 = \bar{X} - b_1 \bar{Y}$$

b_1 is the coefficient of regression of X on Y. Notice also that although there might be no doubt as to which of two variables is the dependent one there could be occasions when the relationship is needed the other way round. For example, the proportion of mice affected in an experiment is dependent on the strength of dose administered, but we might want to know what strength of dose affects 75 per cent of mice.

The regression line for the regression of X on Y is only the same as the regression line for the regression of Y on X when all the points in the sample lie exactly on a straight line. The angle between the two regression lines measures the correlation between X and Y in the sense that the smaller the angle, the stronger the correlation. The measure of correlation most often used is described below.

The coefficient of correlation

The coefficient described is Pearson's product-moment coefficient of correlation (Karl Pearson, 1857–1936) and is derived by considering how well the fitted regression accounts for variation in the data.

Suppose that, as before, X and Y are linearly related and Y is the dependent variable. Variation in X will cause variation in Y because of the dependence of Y on X, and this is accounted for in fitting the linear regression of Y on X. There may also be variation in Y due to other influences.

If most of the total variation in Y is due to variation in X the regression line is a good fit to the relation between X and Y and there is a strong correlation between X and Y.

The *total variation* in Y is $\Sigma(Y_i - \bar{Y})^2$ (compare the expression for the variance of Y). The linear relation fitted gives an estimated value $\hat{Y}_i$ corresponding to X_i instead of the actual value Y_i, so the variation in Y '*explained*' by the linear relation is $\Sigma(\hat{Y}_i - \bar{Y})^2$ (in effect the regression replaces Y_i by $\hat{Y}_i$). The quantity

$$r^2 = \frac{\text{explained variation}}{\text{total variation}}$$

$$= \frac{\Sigma(\hat{Y}_i - \bar{Y})^2}{\Sigma(Y_i - \bar{Y})^2}$$

is used as a measure of the closeness of the points to a straight line and is called a measure of 'goodness of fit'. Thus, if $r^2 = 0{\cdot}9$ for example, variation in X accounts for 90 per cent of the variation in Y and supports an assertion of strong linear dependence of Y on X. A low value of r^2 suggests that a linear relation is inappropriate, as would be shown by a scatter diagram. The highest value of r^2 is $+1$ and in this case every estimated value $\hat{Y}_i$ coincides with the actual value Y_i, for all points (X_i, Y_i) lie on the regression line and the linear relation between X and Y is exact. The lowest value of r^2 is 0 and occurs when there is no relation between X and Y. We can show algebraically that r^2 is the product of the two regression coefficients obtained from the regression of Y on X and from the regression of X on Y.

We call r the *coefficient of correlation* between X and Y and it has the same value as the coefficient of correlation between x and y. We take the positive square root of r^2 when the regression line is of positive slope and the negative square root when it is of negative slope. There is no need to find all the estimated values $\hat{Y}_i$ in order to find r (or r^2) for algebraically we can show that

$$r = \frac{\Sigma x_i y_i}{\Sigma x_i^2 \Sigma y_i^2}$$

or in the form for computation

$$r = \frac{n\Sigma X_i Y_i - \Sigma X_i \Sigma Y_i}{\sqrt{[n\Sigma X_i^2 - (\Sigma X_i)^2]}\sqrt{[n\Sigma Y_i^2 - (\Sigma Y_i)^2]}}$$

The correct sign of r is given automatically by the numerator of this formula, so both the square roots in the denominator are taken as positive. In fact the numerator is the same as the numerator in the expression for b_0 (and that for b_1) and r, b_0, and b_1 all have the same sign.

Referring back to the example given earlier in this chapter on the relation between the amount of time students study each day (X), and the number of cigarettes they smoke (Y), we find that the coefficient of correlation between X and Y is

$$r = \frac{(8)(618{\cdot}8) - (43{\cdot}6)(110)}{\sqrt{[(8)(248{\cdot}24) - (43{\cdot}6)^2]}\sqrt{[(8)(1550) - (110)^2]}}$$

$$= \frac{154{\cdot}4}{\sqrt{(84{\cdot}96)}\sqrt{(300)}}$$

$$= 0{\cdot}967 \text{ (using logarithm tables)}$$

Thus there is a strong positive linear correlation between X and Y.

$$r^2 = 0{\cdot}935 \text{ (using logarithm tables)}$$

so we say that 93·5 per cent of the variation in Y is explained by the regression.

Notice that a coefficient of correlation might be high because of the way the numbers are related to one another rather than because there is a real relation between the variables. This is called a *spurious correlation*. Thus a high coefficient does not enable us to conclude that the variables are in any way related (though if they are linearly related there will be a reasonably strong correlation between them). We must consider the sense as well as the numbers. An example of spurious correlation which is often quoted is a high correlation between the number of births (human) and the number of storks observed in different villages.

Estimation of relationships in the population

As stated earlier in this chapter, a linear regression

$$Y = a_0 + b_0 X$$

fitted for a set of sample data can be considered to be an estimate of a linear relation in the population. Suppose that the linear relation in the population is

$$Y = \alpha + \beta X$$

Then a_0 is an estimate of α and b_0 is an estimate of β.

In order to discuss the sampling distributions (see chapter 6) and properties of the least squares estimators a and b (which have the values a_0 and b_0 for one particular sample) we need to consider the sampling process involved. When regressing Y on X we are finding average Y for fixed X, so when taking other samples of size n we keep the same n values of X in each sample. In the complete population of (X, Y) values there will be several values of Y with each of the n values X_i of X (see Figure 7.3). In sampling we choose for each of the n values X_i one Y value, Y_j, at random from the set of Y values in the population for which $X = X_i$.

Now the value of Y on the population line corresponding to the value X_i of X is $\alpha + \beta X_i$, and the values Y_j which

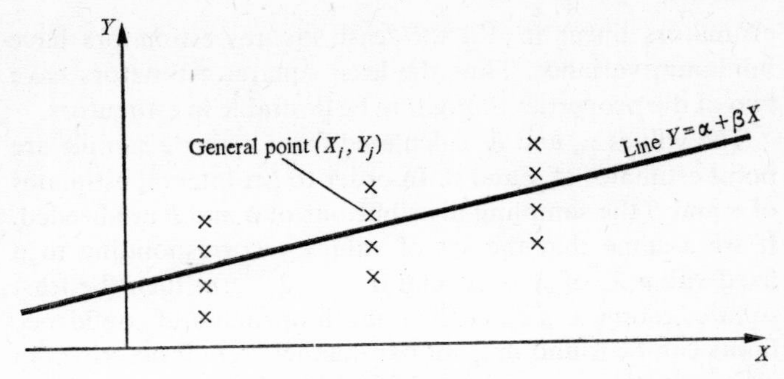

Figure 7.3

actually occur with X_i will be scattered around this value, so we can say

$$Y_j = \alpha + \beta X_i + \epsilon_j$$

where ϵ_j is a random element which we can think of as a disturbance term causing Y_j to take a different value from that given by the exact relationship. We assume that the set of ϵ_j occurring for a fixed value X_i of X is such that the mean value of the ϵ_j is zero and that their variance is σ^2. This holds for each of the n values X_i. Further, the method of sampling ensures that the different ϵ_j are independent of one another, whether arising from the same or different values X_i of X.

With these assumptions it can be shown that

$$E(a) \equiv \text{mean value of } a = \alpha$$

and

$$E(b) \equiv \text{mean value of } b = \beta$$

that is, that the least squares estimators a and b are unbiased estimators of α and β respectively. In addition it can be shown that of all unbiased estimators of α and β of the form $\Sigma w_i Y_i$, where the w_i are constants and the Y_i are the values of Y in the sample used to estimate α and β (i.e.

estimators linear in Y_i) the least squares estimators have minimum variance. Thus the least squares estimators have two of the properties thought to be desirable in estimators.

The values a_0 and b_0 calculated from a single sample are point estimates of α and β. In order to get interval estimates of α and β the sampling distributions of a and b are needed. If we assume that the set of values ϵ corresponding to a fixed value X_i of X is normal ($i = 1, 2, ..., n$) then the least squares estimators a and b are both normal and confidence limits can be found in a similar manner to that described in chapter 6. However, these limits involve σ^2 whose value is usually not known. If σ^2 is estimated from the sample of n pairs (X_i, Y_i) confidence limits for α and β can be found using the distribution known as the t distribution (see Hoel, 1971; Walpole, 1974; Wonnacott and Wonnacott, 1972; Yeomans, 1968). Knowledge of the distributions of a and b also enables us to test hypotheses about the values of α and β.

The coefficient of correlation, r, calculated from the sample is an estimate of the coefficient of correlation between X and Y in the population. If the population of X is assumed to be normal, and the population of Y is assumed to be normal, the distribution of r can be found fairly easily and can be used to give interval estimates of the population coefficient and to test hypotheses about its value (see Hoel, 1971; Yeomans, 1968).

Prediction

If Y depends on X and we have a regression equation estimating the relationship between X and Y then we can use this to find what values of Y occur on the regression line or curve with different values of X, and we call these *predicted* values of Y. However, since the information about the relationship comes from sample data we know nothing about the relationship outside the range of sample data and it is therefore only valid to predict values of Y for values of X in the range of sample values of X (unless we have strong prior reason for supposing the form of the relation to continue outside the range). In other words it is not valid to *extrapolate* (extend) the relationship outside the range of sample values of X.

Different samples give rise to different estimates of the relationship between X and Y and hence different predicted values of Y for the same value of X. For X fixed at X_0 say, the distribution of the predicted values of Y for $X = X_0$ can be found from the distributions of the least squares estimators. The variance of the predicted value of Y for $X = X_0$ gets larger as the distance of X_0 from $\bar{X}$ (the mean of the X values in the sample) gets larger.

Other regressions

This chapter has concentrated on the simplest kind of regression, bivariate linear regression, but the principle of least squares can also be used to estimate bivariate relations of the general form $Y =$ function of X, and for regressions involving more than one explanatory variable. Estimates are obtained by minimizing the sum of squares of deviations from the regression curve or surface, much as in the bivariate linear case. In the case where there are several explanatory variables we can estimate how much variation in the dependent variable is caused by each, and can look at relationships between pairs of variables by holding the other variables at fixed values.

Regression methods are extremely useful in describing and analysing multivariate data and the theory of regression has been well developed. The economist, and in particular the econometrician, uses regression models a great deal, in building models of the UK economy, for instance. Other applications are in agriculture, describing how a yield of crops depends on other variables, and in forecasting, predicting how much electric power will be needed during a cold spell say, or predicting how many hospital beds will be needed in a new town.

Many of the topics introduced in previous chapters are used or taken account of in regression theory, but these topics are useful tools in themselves and lead to the development of many other branches of statistics, as the reader who refers to texts in the list of references will discover.

Exercises

1 Fit the regression $Y = a+bX$ for the data below and calculate the coefficient of correlation between X and Y.

X	-3	-2	-1	0	1	2	3
Y	-20	-12	-3	2	14	16	25

2 Plot a scatter diagram of the data below and use the principle of least squares to estimate the consumption function $C = a+bY$ for a household where C is the household's annual expenditure in £s, Y is its annual disposable income in £s, and observations of C and Y in 6 different years are

Y	1 000	2 000	3 000	5 000	9 000	10 000
C	1 500	2 370	3 300	4 830	8 000	8 740

3 A sample of 15 secondary schools with between 300 and 1 500 pupils is taken in a certain county. Let X denote the number of pupils in a school and Y the number of teaching staff. The values of X and Y in the sample are such that

$$\Sigma X = 12\,300 \qquad \Sigma Y = 417$$

$$\Sigma X^2 = 11\,730\,000 \quad \Sigma Y^2 = 13\,687 \quad \Sigma XY = 399\,600$$

Fit the regression line $Y = a+bX$. Predict the value of Y for a school in the county with 700 pupils (as a point estimate).

4 Eight pairs of observations on X_1 and X_2 are such that

$$\Sigma X_1 = 4 \qquad \Sigma X_2 = 15$$
$$\Sigma X_1^2 = 44 \qquad \Sigma X_2^2 = 175 \qquad \Sigma X_1 X_2 = 85$$

Calculate the regression coefficients for the two linear regressions – X_1 on X_2, and X_2 on X_1. Find the product of these two coefficients. What does it tell you about the relationship between X_1 and X_2 in this case?

5 The following information is obtained for 5 different men: X = number minutes' walk man lives from station, and Y = number of minutes he has to wait for a train on average.

Calculate the coefficient of correlation between X and Y.

X	1·8	2·0	2·8	3·7	4·2
Y	5·0	3·5	2·5	2·0	0·5

6 In a quality control process two measurements are made, X_1 which is cheap to measure, and X_2 which is costly to measure. A sample of measurements on 9 items is given below.

Does it seem reasonable to suggest measuring X_1 alone? Justify your answer diagrammatically and by calculation.

X_1	1·0	1·3	0·9	1·1	1·5	1·4	1·2	1·1	1·3
X_2	2·4	3·4	1·4	2·2	3·0	3·0	3·2	2·1	2·9

7 The marks obtained by 8 students in examinations in Economics and Mathematics are shown below. Represent the marks on a scatter diagram. Calculate the coefficient of correlation between them.

Economics	50	45	55	60	50	40	53	50
Mathematics	64	60	70	72	65	53	70	60

8 Find the estimated values of Y on the least squares regression line $Y = a+bX$ for the data below and use them to find the variation explained by the regression. Find also the total variation of Y and hence the coefficient of correlation between X and Y.

X	-3	-2	0	1	4
Y	3	2	2	1	-3

Appendix 1

Recommended basic statistical sources for community use

Two lists, List I, The Basic Minimum List, and List II, The Extended List, have been drawn up by a joint working party of librarians and economic statisticians set up by the Library Association and the Royal Statistical Society.

Titles in List I are:

1 *Abstract of Regional Statistics* (annually)
2 *Annual Abstract of Statistics*
3 *Trade and Industry* (weekly) (formerly *Board of Trade Journal*)
4 *Monthly Digest of Statistics*
5 *Statistics on Incomes, Prices, Employment and Production* (quarterly)
6 *United Nations Statistical Yearbook*

For libraries in Scotland:

7 *Digest of Scottish Statistics* (two issues a year)

For libraries in Wales:

8 *Digest of Welsh Statistics* (annually)

For libraries in Northern Ireland:

9 *Digest of Statistics for Northern Ireland* (two issues a year)

APPENDIX 1

Full details and List II are published in *Journal of the Royal Statistical Society*, Series A, vol. 132, Part 1, 1969.

Appendix 2

The summation sign $\sum$ (pronounced sigma)

ΣX (sigma X) means add together all the different values of X, so if X takes values 2, 0, -3, 4, and 10

$$\begin{aligned}\Sigma X &= 2+0+(-3)+4+10\\ &= 16-3 = 13\end{aligned}$$

Sometimes it is convenient to denote the different values of X by $X_1, X_2, \ldots, X_n$ so we can then say

$$\sum_{i=1}^{n} X_i \text{ or } \Sigma X_i = X_1+X_2+\ldots+X_n$$

Similarly ΣX^2 means add together all the different values of X^2, $\Sigma \log X$ means add together all the different values of $\log X$, ΣXY means add together all the different values of XY, and so on.

Notice that if c is a constant

$$\begin{aligned}\Sigma cX_i &= cX_1+cX_2+\ldots\\ &= c(X_1+X_2+\ldots)\\ &= c\Sigma X_i\end{aligned}$$

Similarly Σ(c times function of X) = $c\Sigma$ (function of X)

If $f(X)$ is the frequency of X

$$\Sigma f(X) = \text{sum of the different frequencies of } X$$
$$= \text{total frequency of } X \text{ (usually denoted by } n\text{)}$$

Noting that $\bar{X}$ is a constant and $\Sigma X f(X) = n\bar{X}$ we now see that the variance of X is

$$\begin{aligned}\Sigma(X-\bar{X})^2 f(X) &= \Sigma(X^2-2\bar{X}\,X+\bar{X}^2)f(X)\\ &= \Sigma X^2 f(X)-2\bar{X}\,\Sigma X f(X)+\bar{X}^2\ \Sigma f(X\\ &= \Sigma X^2 f(X)-2n\bar{X}^2+n\bar{X}^2\\ &= \Sigma X^2 f(X)-n\bar{X}^2\end{aligned}$$

Solutions

Chapter 2

1 i Ratio, discrete; ii Ratio, continuous
iii Ordinal; iv Ratio, discrete
v Ordinal; vi Nominal
vii Ratio, discrete; viii Ordinal
ix Ratio, continuous; x Nominal
xi Nominal; xii Interval, continuous

2

		Week-days (Mons–Thurs)	*Fridays*	*Saturdays*	*Sundays*
Air	Europe	30	75	60	42
	Elsewhere	15	10	9	7
Ship	Europe	21	39	39	18
	Elsewhere	15	20	33	29

3 Column totals: 7 600; 6 600; 4 800; 600; 400
Row totals: 5 200; 2 000; 5 600; 7 200
Grand total: 20 000

In table below R = percentage described under (*a*); C = percentage described under (*b*); T = percentage described under (*c*).

R	38	33	24	3	2	
C	26	26	26	26	26	
T	9·88	8·58	6·24	0·78	0·52	26
R	34·2	36·3	20·4	5·7	3·4	
C	9	11	8·5	19	17	
T	3·42	3·63	2·04	0·57	0·34	10
R	33·25	24·74	33	5·25	3·75	
C	24·5	21	38·5	49	52·5	
T	9·31	6·98	9·24	1·47	1·05	28
R	42·75	38·5	18	0·5	0·25	
C	40·5	42	27	6	4·5	
T	15·39	13·86	6·48	0·18	0·09	36
T	38	33	24	3	2	100

5

No. children in family	*Frequency*	*Relative frequency*	*Cumulative frequencies*	
1	4	0·08	4 < 2	1 ≥ 9
2	16	0·32	20 < 3	2 ≥ 8
3	9	0·18	29	4
4	7	0·14	36	5
5	9	0·18	45	14
6	1	0·02	46	21
7	2	0·04	48	30
8	1	0·02	49	46
9	1	0·02	50	50
	50	1·00		

Appropriate diagram is line chart.

6

No. employees eating	*Frequency*	*Relative frequency*	*Cumulative frequencies*	
57	13	0·19	$13 < 58$	$6 \geq 64$
58	14	0·20	$27 < 59$	$16 \geq 63$
59	8	0·11	35	25
60	4	0·06	39	31
61	6	0·09	45	35
62	9	0·13	54	43
63	10	0·14	64	57
64	6	0·09	70	70
	70	1·01*		

* Sum > 1 due to rounding.

Appropriate diagram is line chart.

7

No. of questions	*Frequency*	*Mid-points*	*Cumulative frequencies*	
0–9	1	4·5	$1 < 10$	$3 \geq 90$
10–19	2	14·5	$3 < 20$	$7 \geq 80$
20–29	5	24·5	8	16
30–39	4	34·5	12	26
40–49	10	44·5	22	38
50–59	12	54·5	34	48
60–69	10	64·5	44	52
70–79	9	74·5	53	57
80–89	4	84·5	57	59
90–99	3	94·5	60	60
	60	(Q. asks for one only)		

Common width of classes is 10. The variable is discrete. This is best grouping here as all values which could occur are allowed for. Other labellings of classes are acceptable.

8 Close two open-ended classes – say make class 'under 7' class '0–7' (nobody earns less than nothing); and make class '40 and over' class '40–45' (nature of distribution suggests that small percentage of men with incomes in this group do not earn much more than £40 a week).

Mid-points	*Widths*	*Cumulative distributions*	
3·5	7	0·5 < 7	0·2 ≥ 40
9·0	4	18·1 < 11	0·6 ≥ 30
13·0	4	58·0	4·8
17·0	4	84·4	15·6
21·0	4	95·2	42·0
26·5	7	99·4	81·9
35·0	10	99·8	99·5
42·5	5	100·0	100·0
		(Q. asks for one only)	

Notice that the variable (income) is discrete, but has to be treated as continuous here. Take care when drawing the histogram since the classes are not all the same width.

9

Mid-points	*Cumulative frequencies (one needed for ogive)*	
17·5	0·21 < 20	0·03 ≥ 60
22·5	0·37 < 25	0·08 ≥ 55
27·5	0·45	0·15
32·5	0·53	0·25
37·5	0·63	0·37
42·5	0·75	0·47
47·5	0·85	0·55
52·5	0·92	0·63
57·5	0·97	0·79
62·5	1·00	1·00

All classes are of width 5. The variable is continuous.

10 Appropriate diagrams are:

a Bar chart, or pictogram, or pie chart
b Time series
c Histogram (close first and last classes)
d Bar chart
e Log graph shows rates of change very similar (time series does not).

Chapter 3

1

	Mode	*Median*	*A.M.*	*Range*
Chapter 2, q. 5	2	3	3·46	1 to 9
Chapter 2, q. 6	58	59½	60·04	57 to 64

2 Mean = 6·294 Median = 6 Mode = 6
Accounts department should take some account of variability in data.

3 ii 56⅔. From grouped distribution, 58 exactly
iii Approx. 15 iv 55½ (from grouped distribution)

4 Modes at 4·8 and 43·25 Median = 24·5 A.M. = 23·51
The modes reflect the U-shape of the distribution.
S.D. = 17·18 Semi-interquartile range = 17·49

5 Median = 14·20 Semi-interquartile range = 2·94

6 i 63·2 ii Median = 59·76,
semi-interquartile range = 7·89
iii (*a*) 87 (*b*) 68

7

	A.M.	*S.D.*	*C. of V.*
Subject 1	53·65	6·75	0·1258
Subject 2	56·69	7·29	0·1285

8

	Shape	*A.M.*	*Variance*
A	+ve skew	25·5	254·75
B	−ve skew	54·3	256·51

Chapter 4

1 i $\left(\frac{3}{6}\right)\left(\frac{3}{6}\right) = \frac{1}{4}$ ii $\left(\frac{2}{6}\right)\left(\frac{1}{6}\right) = \frac{1}{18}$

iii $\left(\frac{4}{6}\right)\left(\frac{1}{6}\right) = \frac{1}{9}$

2

i $\frac{1}{4}$ ii $\frac{1}{13}$ iii $\frac{1}{52}$ iv $\frac{1}{13}$

v $\frac{4}{52}+\frac{13}{52} = \frac{17}{52}$ vi $\frac{1}{52}+\frac{1}{52} = \frac{1}{26}$ vii $\frac{1}{26}$

viii $\frac{1}{4}+\frac{1}{4} = \frac{1}{2}$ ix $\frac{1}{13}+\frac{1}{13} = \frac{2}{13}$

3

i All tosses are tails. Prob $= \left(\frac{1}{2}\right)\left(\frac{1}{2}\right)\left(\frac{1}{2}\right) = \frac{1}{8}$

ii 1 − prob (no toss is heads) $= 1-\frac{1}{8} = \frac{7}{8}$

iii Head occurs either on 1st toss or 2nd or 3rd, tails on other two tosses. Prob $= \frac{1}{8}+\frac{1}{8}+\frac{1}{8} = \frac{3}{8}$

iv Exactly two heads means exactly one tail $\therefore$ prob $= \frac{3}{8}+\frac{3}{8} = \frac{3}{4}$

v $\frac{1}{2}$

vi $\left(\frac{1}{2}\right)\left(\frac{1}{2}\right) = \frac{1}{4}$

vii prob (head on first) + prob (tail on last) − prob (head on first and tail on last) $= \frac{1}{2}+\frac{1}{2}-\frac{1}{4} = \frac{3}{4}$

viii $\frac{1}{2}$

4

Prob (green on 1st and green on 2nd) + prob (black on 1st and green on 2nd)

$$= \left(\frac{5}{8}\right)\left(\frac{7}{10}\right)+\left(\frac{3}{8}\right)\left(\frac{5}{7}\right) = \frac{79}{112}$$

5

	Experiment 1	*Experiment 2*
(*a*)	$\frac{1}{4}$	$\frac{1}{3}$
(*b*)	$\frac{1}{4}$	0

(*c*) $\frac{1}{4}$ Either 1st is *A* or *B* or *C* or *D* and with each first draw *B* is 2nd

$$\text{Prob} = 3 \times \frac{1}{4} \times \frac{1}{3} = \frac{1}{4}$$

6 Use a Venn diagram.
i 0·07 ii 0·403 iii 0·482

7 Use a Venn diagram.
0, 0·05, 0·35, 0·5

8 (*a*) $\frac{\text{No. children in families of 4 children}}{\text{Total no. children}} = \frac{28}{173}$

(*b*) $\frac{\text{No. families with 4 children}}{\text{Total no. families}} = \frac{7}{50}$

9 3 to 2

10 i $\frac{1}{2}$

ii By Bayes's Theorem $= \frac{0{\cdot}06}{0{\cdot}06+0{\cdot}025+0{\cdot}0375}$

$= 0{\cdot}48978$

iii $0{\cdot}06+0{\cdot}025+0{\cdot}0375 = 0{\cdot}1225$

Chapter 5

1 i 3/8 ii 0·4 iii 0 iv 0·20 v cannot be distribution

2 i

X	$f(X)$
0	1/8
1	3/8
2	3/8
3	1/8

ii

X	$f(X)$	X	$f(X)$
2	1/36	8	5/36
3	2/36	9	4/36
4	3/36	10	3/36
5	4/36	11	2/36
6	5/36	12	1/36
7	6/36		

3 $c = 1/15$

X	-2	-1	0	1	2
$f(X)$	1/15	4/15	5/15	4/15	1/15

$\text{var}(X) = 16/15$

Diagram is line chart

4 i

X	$f(X)$
0	0·343
1	0·441
2	0·189
3	0·027

Mean = 0·9
Variance = 0·63

ii

X	$f(X)$
0	0·0625
1	0·2500
2	0·3750
3	0·2500
4	0·0625

Mean = 2·0
Variance = 1·0

5 Binomial
Prob (rings 0, 1, or 2) = $(1+15+90)/4^5 = 53/512$

6 Negative binomial
i Prob = 0·2592 ii Prob = 0·20736
Average no. = 5

7 i 0·3497 ii 0·1359 iii 0·4603 iv 0·1093
v 0·3936

8 i 0·8622 ii 221 grams and over

9 8·34% 56·45 8·68%

10 Normal approximation to binomial
i 0·0475 ii 0·3779 iii 0·3085

Chapter 6

1 For example:
- i Heights of all trees, of all oak trees, of trees in London
- ii members of societies, British persons, members of this society at any time
- iii salaries of teachers, of professional people, of nursery school teachers at any time.

2 Mean of population = 5/2
(*a*) Samples are (1,2), (1,3), (1,4), (2,3), (2,4), (3,4)
(*b*) Samples as in (i) plus (1,1), (2,2), (3,3), (4,4)

3 i 733·96 to 736·04 ii 8·674 to 11·326 iii 0·510 to 0·574

4 i 4·9412 to 5·0588 ii 4·1613 to 4·2387

5

i	*Given place*	*Not given place*
Has qualifications		Type I error
Does not have qualifications	Type II error	

ii	*Drug prescribed*	*Drug not prescribed*
Has disease X		Type I error
Does not have disease X	Type II error	

6 Critical z at 5 per cent level, two-sided test are $\pm 1{\cdot}96$. Sample $z = 3{\cdot}66$ ∴ reject hypothesis that average is 3·18.

7 Yes. Critical z at 5 per cent level, one-sided test is 1·64. Sample $z = 2{\cdot}032$

8 Two-sided test. Sample $z = 0{\cdot}617$. Engineer A accepts no difference between means at 1 per cent, 5 per cent, 10 per cent levels of significance. Engineer B bases a test on the t distribution.

9 Critical z at 10 per cent level, 2-sided test are $\pm 1{\cdot}64$. Critical values of p are 0·4331 and 0·5669 ∴ accept $p = 0{\cdot}5$.

10 i One-sided test. Sample $z = 2{\cdot}545$
Prob value = 0·0055
ii Two-sided test. Sample $z = -1$
Prob value = 0·3174

Chapter 7

1 $Y = 7{\cdot}43X + 3{\cdot}14$ $\quad r = 0{\cdot}994$

2 $C = 790{\cdot}7 + 0{\cdot}8Y$

3 $Y = -0{\cdot}96 + 0{\cdot}035X$ $\quad Y = 23{\cdot}54$

4 124/235 for X_1 on X_2, 155/84 for X_2 on X_1
$r^2 = 961/987$ – close to 1 suggesting strong linear relation between X_1 and X_2

5 $r = -0{\cdot}947$

6 $r = 0{\cdot}804$ Fairly high correlation, but not so high that measuring X_1 alone is much the same as measuring both X_1 and X_2. We should also need to consider the relative costs and importance of X_1 and X_2 before reaching a decision.

7 $r = 0{\cdot}941$

8 $Y = 1-0{\cdot}8X$
Predicted values of Y are 3·4, 2·6, 1, 0·2, −2·2
Explained variation = 19·2 Total variation = 22·0
$r = \sqrt{(19{\cdot}2/22)} = -0{\cdot}934$ (same sign as regression coefficient).

References

AITCHISON, J. and BROWN, J. A. C. (1969), *The Lognormal Distribution with Special Reference to its Uses in Economics*, Cambridge University Press.

BARTHOLEMEW, D. J. (1967), *Stochastic Models for Social Processes*, Wiley, Chichester and New York.

BUGG, D. D., HENDERSON, M. A., HOLDEN, K., and LUND, P. J. (1968), *Statistical Methods in the Social Sciences*, North-Holland Publishing Company, Amsterdam.

CENTRAL STATISTICAL OFFICE (1972), *Facts in Focus*, Penguin Books, Harmondsworth, in association with HMSO, London.

DAVID, F. N. (1962), *Games, Gods and Gambling*, Griffin, London.

DEPARTMENT OF EMPLOYMENT (annual), *Family Expenditure Survey*, HMSO, London.

GOVERNMENT STATISTICAL SERVICE (1971), *Social Trends No. 2*, HMSO, London.

HOEL, P. G. (1971), *Elementary Statistics*, 3rd edition, Wiley, Chichester.

HUFF, D. (1954), *How to Lie with Statistics*, Gollancz, London.

KALTON, G. (1966), *Introduction to Statistical Ideas for Social Scientists*, Chapman & Hall, London.

KEMENY, J. G., SNELL, J. L., and THOMPSON, G. L. (1957), *Introduction to Finite Mathematics*, Prentice-Hall, Englewood Cliffs, New Jersey.

LINDLEY, D. V. and MILLER, J. C. P. (1966), *Cambridge Elementary Statistical Tables*, Cambridge University Press.

MOSER, C. A. and KALTON, G. (1971), *Survey Methods in Social Investigation*, Heinemann, London.

OFFICE OF POPULATION CENSUSES AND SURVEYS (1961, 1966, 1971), *Census of population reports*, HMSO, London.

WALPOLE, R. E. (1974), *Introduction to Statistics*, 2nd edn, Collier-Macmillan, London.

WONNACOTT, T. H. and WONNACOTT, R. J. (1972), *Introductory Statistics for Business and Economics*, Wiley, Chichester and New York.

YEOMANS, K. A. (1968), *Introductory Statistics, Statistics for the Social Scientist*, 2 vols, Penguin Books, Harmondsworth.

ZEISEL, H. (1958), *Say it with Figures*, Routledge & Kegan Paul, London

4, 8, 12, 16, 20, 50

$\frac{110}{6} = \frac{17}{4}$

median = 4

mid-rang = $\frac{50+4}{2}$

= 27

$\frac{84}{6}$ [14] $\bar{x}$

Median = 14

Mid range = 14

4, 8, 12, 16, 20, 24

4 + 24 = $\frac{28}{2}$